Adrian Kind
How Does the Psychiatrist Know?

Bioethics / Medical Ethics | Volume 7

Adrian Kind (Dr.) works as a postdoctoral researcher at Charité – Universitätsmedizin Berlin and is currently training as a psychodynamic psychotherapist. He received his doctorate through the partner project "Extrospection", a collaboration between Humboldt-Universität zu Berlin and Otto-von-Guericke-Universität Magdeburg. He holds Master's degrees in philosophy (Humboldt-Universität zu Berlin) and psychology (Internationale Psychoanalytische Universität Berlin). His research focuses on epistemological and metaphysical questions in the philosophy of psychiatry, philosophy of mind, philosophy of sports, and the philosophy of religion.

Adrian Kind

How Does the Psychiatrist Know?

On the Epistemology of Psychiatric Diagnostic Reasoning

[transcript]

The publication of this work was supported by the Open Access Publication Fund of Humboldt-Universität zu Berlin.
The publication of this work was supported by the Open Access Publication Fund of Otto von Guericke University of Magdeburg.
The research was funded by the Deutsche Forschungsgemeinschaft (DFG, German Research Foundation) – 337619223.

HUMBOLDT-UNIVERSITÄT ZU BERLIN

Funded by DFG Deutsche Forschungsgemeinschaft German Research Foundation

Bibliographic information published by the Deutsche Nationalbibliothek
The Deutsche Nationalbibliothek lists this publication in the Deutsche Nationalbibliografie; detailed bibliographic data are available in the Internet at https://dnb.dnb.de/

transcript Verlag | Hermannstraße 26 | D-33602 Bielefeld | live@transcript-verlag.de

Cover layout: Maria Arndt
Cover illustration: The cover image was created using Adobe AI in InDesign.
Printed by: Elanders Waiblingen GmbH, Waiblingen
https://doi.org/10.14361/9783839476741
Print-ISBN: 978-3-8376-7674-7
PDF-ISBN: 978-3-8394-7674-1
ISSN of series: 2702-8267
eISSN of series: 2702-8275

Printed on permanent acid-free text paper.

Contents

Acknowledgments

This book is based on my PhD thesis, written during my time in the DFG-funded Research Training Group Extrospection (GRK 2386) at the Berlin School of Mind and Brain, Humboldt-Universität zu Berlin, and the Section of Philosophy at Otto-von-Guericke Universität Magdeburg. Its publication has been made possible through the support of the open-access publication funds of both university libraries. I am deeply grateful to my supervisors, Sascha B. Fink and Henrik Walter, for their guidance throughout this project. I also thank my colleagues, whose insights and support have been invaluable. Special thanks go to those who inspired me to pursue a university education and my PhD: Monika Pohlmann, Markus Tiedemann, Bernd Ruffer, and Lili Gast. I also wish to acknowledge the influence of longstanding companions such as Mattis, Erik, Edwin, Gerry, Stefan, and Sera, as well as new friends like Lukas, Marlo, and Trevor, who helped me endure the ups and downs along the way. Most importantly, I thank my family for their unwavering support, and Marlene for her constant encouragement and trust in me.

0. Introduction

If we want to know whether someone is suffering from a mental disorder, we send them to a psychiatrist. We do so apparently because we believe that psychiatrists enjoy epistemic superiority when it comes to diagnosing mental disorders, relative to non-experts in the field of psychiatry. Although it is *prima facie* plausible to assume that psychiatrists have a better capacity for diagnostic judgement than individuals untrained in the field of psychiatry, this assumption itself raises questions. One of these questions, the one I will be concerned with in this part of the thesis, is: how do psychiatrists arrive at their diagnostic conclusions?

This question is of importance to philosophy of psychiatry and should also be of interest to clinicians themselves. It deserves philosophical attention because answering it is a requirement for a systematic understanding of the epistemology of psychiatry, which consists not only of epistemic issues around the psychiatric sciences and the choice of medical interventions, but also of diagnostic decision-making. Moreover, developing an understanding of psychiatric diagnostics is a requirement for enabling us to discuss other phenomena of ethical and epistemological interest in psychiatry. These phenomena include the ethically important task of deciding when a diagnostic decision is just wrong and when it is malpractice, and how to understand the social-epistemological dynamics involved in resolving expert disagreements regarding diagnosis. Addressing these and other topics are desiderata for a theory of psychiatric diagnostics that can be addressed in meaningful depth only on the foundation of an established understanding of the diagnostic process itself. The three major aims of this thesis are: (1) to provide a new proposal for how psychiatrists arrive at their diagnostic judgements, (2) demonstrate how this proposal enables us to address several desiderata of a philosophical account to psychiatric diagnostics, and (3) to defend this approach against existing alternative approaches is the aim of this book.

Before the real work begins, I will use this Introduction to set the scene. I will (0.1) reformulate and clarify the causal question "How do psychiatrists arrive at their diagnostic conclusions?" to prepare it for a philosophical treatment. After that I will (0.2) foreshadow the answer to the Methodological Question that I will develop and answer in this thesis. Finally, (0.3) I will lay out the chapter structure of the thesis and (0.4) make the transition to Chapter 1.

0.1 The Methodological Question

In order to develop a philosophical account that can answer the question "how do psychiatrists arrive at their diagnostic conclusions?", we need to some preparatory work on the question itself. This preparation will clarify how I think the question should be understood for the purpose of a philosophical investigation and thus what to expect from an adequate answer to it. Doing so will avoid misunderstandings regarding my project. First, therefore, let me concretise how I understand the question.

I will take the question of how psychiatrists arrive at their diagnostic judgements to be a question about the *method* used by psychiatrists to make their diagnostic judgements. Why a method? According to Goldman (2000), learned belief-forming procedures shape our inquiry. To think of diagnostics carried out by psychiatrists as following a learned belief-forming procedure to arrive at their diagnostic conclusions seems *prima facie* plausible given that psychiatrists are medical experts who receive scientific and clinical education acquiring knowledge and skill for their clinical work, including diagnostics. Plausibly, they are not born with diagnostic insight; they learn what to do to generate it. If we consider psychiatric diagnostics to be a method, asking how it works is about asking questions of methodology. Thus, I will call the question I work towards answering in this book the Methodological Question. "What is the method of psychiatric diagnostics?" Next, let me lay out what will be required in order to answer the Methodological Question – that is, in order to propose a clinical methodology of psychiatric diagnostics.

Providing an answer to the Methodological Question has adequacy conditions and desiderata. The adequacy conditions are the minimal requirements a proposed answer should meet to provide a proper answer to the Methodological Question. The desiderata are things we want from the adequate answer to the Methodological Question to make it an actually good answer; they are factors that, if a given answer offers more of them than another, might make this answer preferable to others. I will discuss both aspects in turn.

The adequacy conditions for an answer to the Methodological Question derive from the question's format. Because we are asking about the methods of diagnostic reasoning, it is a Methodological Question. And, again according to Goldman (2000), methodologies are theories of methods that, as such, describe, explain, and evaluate methods of inquiry. To address the Methodological Question, given Goldman's characterisation, and to provide a theory of the method of psychiatric diagnostic reasoning, we need to do three things:

1. We need to adequately describe the method at work behind the diagnostic process. What does this method look like? How does it operate? When are its constituent steps carried out?
2. We need to explain the rationale behind the method. What purpose do the steps in the method serve? How are these steps thought to contribute to the achievement of the epistemic end of the method used?
3. We need to lay out what to think of the justificatory status of beliefs achieved using this method. How are specific aspects of the method thought to justify its outcomes? Can we say something general about how promising the method is for arriving at true conclusions, or at least set out how we may make such judgements for specific instances of the methods used?

Beyond these general adequacy conditions that provide the minimal requirements for an answer to the Methodological Question, there are some intuitive desiderata for an answer. While the adequacy conditions stated in the last paragraph derive purely from the Methodological Question requiring a methodological proposal, additional desiderata derive from its target: psychiatric diagnostics. If the proposal wants to do more than meet some minimal criteria – that is, if it aspires to explore some aspects of psychiatric diagnostics in reasonable depth – these desiderata should be met. In what follows, I offer a list of plausible desiderata, some of which were already mentioned when motivating the Methodological Question.

1. An answer to the Methodological Question should allow us to make sense of the relevant steps of the diagnostic process. A proposal should not leave major aspects poorly understood, lacking a rationale for their existence in the process. Only then can we say that the proposal really encompasses psychiatric diagnostics.
2. An answer to the Methodological Question should provide a proposal that is cognitively realistic. By cognitively realistic I mean that the way the proposed method describes psychiatric diagnostics as the activity of psychiatric experts should account for the engagement of psychiatrist in that process in a way that not only is able in principle to make sense of the steps of the diagnostic process (as required in my last point) but does so in a way that appears to be attainable and realistically undertaken by psychiatrists as cognitive agents, if only under ideal circumstances (e.g., no time pressure). The desired proposal does not require psychiatrists to think or act in a way that goes obviously beyond an expert human capacity; rather, it seems to be a plausible intentional cognitive and behavioral procedure carried out by clinicians. This will prevent the answer from being more than a proposal for understanding diagnostics that works in the armchair but bears minimal relation to real practice.

3. The answer to the Methodological Question should allow us to explicate the internal standards embodied in actual diagnostic practices, enabling us to say when failure counts as malpractice or just a wrong diagnosis.[1] The capacity to make such crucial distinctions regarding the outcomes of diagnostic process matters in part because it concerns central legal and ethical issues in diagnostics, but also because the denotes an appropriately deep understanding of diagnostic standards.
4. An answer to the Methodological Question should be able to explain the occurrence and resolution of diagnostic uncertainty, for example as regards whether or not one should attribute a specific symptom or diagnose a specific disorder diagnose in a patient. Accounting for uncertainty and its resolution is an obviously relevant requirement given the frequent day-to-day occurrence of this phenomenon in diagnostic clinical work.
5. An answer to the Methodological Question should make sense of the phenomenon of the sometimes-observed good diagnostic "instincts" of experienced clinicians who rapidly come up with potential diagnostic conclusions and often turn out to be right. The answer to the Methodological Question should enable us to understand how these "instincts" work, and how to assess their conclusions in relation to the internal standards of psychiatric diagnostics. Again, being able to explain and evaluate this phenomenon is relevant given its apparent prevalence in clinical practice and the resulting question of whether or not it is permissible to make instinctive diagnoses.
6. An answer to the Methodological Question should be able to help us understand the occurrence of diagnostic disagreements amongst individual clinicians, as well as amongst the same clinician's judgements over time. Again, this matters because such disagreements are part of everyday clinical reality; being unable to

1 What do I mean by "internal standards"? Internal standards are epistemic norms that psychiatrists ought to follow to arrive at permissible diagnostic conclusions. Conclusions will be considered permissible because they are considered justified by the standards of the expert clinical community that espouses this standard. Although the justification established in this way does not in itself guarantee the desirability of the diagnostic judgements as a function of any "objective" well-groundedness or reliability (this would require further argument), it is nonetheless relevant to matters of responsibility and culpability. Consider the case where a psychiatrist's diagnosis is wrong, but she works in accordance with the internal standards. She will not be considered culpable, as she was justified in making this diagnosis. If, on the other hand, a psychiatrist guesses a diagnosis and thereby violates the internal standards of psychiatric diagnostics, he will be judged culpable of diagnostic malpractice. To make sense of this, both internal standards and the corresponding understanding of justification are relevant. For a similar take on justification, see Pollock (1986, p. 125); Carter and Littlejohn (2021, pp. 320–322.).

address a common phenomenon in diagnostic practice would render an answer of little explanatory use.
7. Finally, it would be desirable that an answer to the Methodological Question could help us understand how changes and hopefully improvements in our understanding of psychopathological conditions are integrated into existing diagnostic practice, or might lead to changes in the framework of psychiatric diagnostics itself. This capacity would make an answer a useful tool for thinking about the (near) future of psychiatric diagnostics.

I have now presented a proposal regarding the Methodological Question and a statement as to what will be required to answer it productively. However, something about the Methodological Question is still vague.

Speaking of a method of psychiatric diagnostics *per se* seems problematic. "Psychiatric diagnostics" considered broadly is an ambiguous term because psychiatric diagnostics is a heterogeneous epistemic practice. Looking at it from a historical perspective, or systematically within any given period of its history, would reveal many methods that physicians concerned with mental disorders have used to diagnose their patients. Accordingly, to provide a presentation of "psychiatric diagnostics" as referring to "everything that every psychiatrist ever did to find out about their patients' psychopathological status" would be an encyclopaedic task. Not only is such a task beyond the scope of what I can do in this project; moreover, it stands to reason that given the heterogeneity of psychiatric diagnostic approaches, considering diagnostics so broadly would doom to failure any attempt to identify a single common method behind all these different ways to diagnose. To avoid this problem, I will limit the scope of my analysis of psychiatric diagnostics – and accordingly of the underlying diagnostic reasoning – to a sufficiently homogenous set of practices to offer a manageable explanandum as a target for the Methodological Question.

For the purpose of answering the Methodological Question, I will consider psychiatric diagnostics to consist of diagnostic efforts carried out by trained professionals through their cognitive and behavioural efforts to arrive at diagnostic conclusions. This process is usually called *clinical diagnostic reasoning*. But I will be even more specific, because this first limitation is still too broad. The diagnostic reasoning of clinicians may vary significantly, and to treat "diagnostic reasoning" as co-extensive with "everything that any psychiatrist ever did to arrive at the diagnostic proposal" is not a promising basis for arriving at a common method and methodology. Hence the kind of diagnostic reasoning I will focus on will be what I understand to be at the heart of (1) *contemporary* and (2) *proper* diagnostic reasoning practice. The question then, of course, is how I determine what I will regard as instances of such contemporary and proper psychiatric diagnostic reasoning.

To gain insight into what constitutes proper contemporary diagnostic reasoning procedures will involve looking at recent authoritative sources on psychiatric diagnostics. I will take into consideration recent position papers and practice guidelines from relevant expert communities such as the *American Psychiatric Association*, as well as recent editions of autorotative textbooks for psychiatric training such as *Kaplan and Sadock's Synopsis of Psychiatry* (Sadock and Sadock, 2014). Setting this focus ensures that my answer to the Methodological Question addresses what is widely accepted as a proper and contemporary approach to psychiatric diagnostic reasoning within the psychiatric expert community itself, and thus that my answer will be of interest to a wide audience. More anachronistic or obscure approaches to diagnostics that deviate from what is widely held to be the state of the art within psychiatry will therefore not be reflected in this work. In sum, the Methodological Question, if we spell out the version that I will work with, is: **what is the method of proper contemporary psychiatric diagnostic reasoning?**

Narrowing our focus to contemporary proper psychiatric diagnostic reasoning leads to a final point concerning the Methodological Question, namely whether it is a normative or a descriptive question and so whether my answer to the Methodological Question should accordingly be consider prescriptive (i.e., normative) or descriptive. The Methodological Question is not a clear-cut example of either a normative or a descriptive question; nor will an answer fall neatly into either of these categories. Rather, both the question and the answer will have to involve both description and normativity. They are descriptive because by inquiring into what the method at work in psychiatric diagnostics is, the question and its answer are concerned with an actual state of affairs that is targeted by the question and can be explained by its answer. The question and answer also have a normative side, since the exemplification of the method at work is supposed to exemplify, specifically, what the *proper contemporary* method is. As such, answering the question will result in a proposal that has the normative force of claiming that one must follow this method if one wants to practise psychiatric diagnostics in accordance with the currently widely shared standards of the clinical psychiatric community. This normativity, however, does not derive directly from any facts of the described method itself (it thus avoids the trap of deriving an *ought* from an *is*); rather, an answer to the Methodological Question gains normative character from the initial normative character of the descriptions of the diagnostic practice on which the proposal of the method is based. Specifically, normativity derives from guidelines and teaching literature intended to say how diagnostics *ought* to take place by establishing relevant standards.

Let me sum up my discussion of the Methodological Question. I plan to address the question of how psychiatrists arrive at their diagnostic conclusions, interpreting this inquiry as what I called the Methodological Question. Namely: What is the method of proper contemporary psychiatric diagnostic reasoning? To address this question adequately, I will present a methodology of psychiatric diagnostic reasoning, providing a description of the method being used, the rationale behind its procedures, and how its conclusions are deemed justified. The resulting methodological proposal will, on the one hand, be descriptive regarding the method in place, but on the other hand, it will have some normative relevance because it is the description of a practice derived from sources that not only present the method but propose that this is the proper method one ought to follow in psychiatric diagnostics if one practises in accordance with the expert community's self-imposed standards for good diagnostic practice.

This proposal will have to meet several requirements. At a minimum, it will have to encompass the necessary aspects of a methodology: a description of the method that I claim to be enacted in psychiatric diagnostic reasoning; an explanation of the rationale behind the method's procedures; and an account of the justifications given to conclusions achieved by these procedures. Moreover, I will have to establish that the descriptive part of my methodological proposal is indeed embodied in diagnostic reasoning practices, to make the presented methodology plausibly apply to psychiatric diagnostics. Beyond this minimal requirement, there are seven desirable features that an answer to the Methodological Question should provide. First, to show a close match between individual aspects of the method and the diagnostic procedure, leaving no aspects of the diagnostic procedure unexplained. Second, to provide a cognitively realistic proposal. Third, to enable the differentiation between misdiagnosis and diagnostic malpractice. Fourth, to explain the occurrence and resolution of diagnostic uncertainty. Fifth, to understand and evaluate diagnostic "instincts". Sixth, to explain diagnostic disagreements and their resolution. And seventh, to enable discussion of the impact that progress in our understanding of psychopathology might have on psychiatric diagnostics.

Now that the Methodological Question is established, and I have discussed what an answer to this question should look like, I will proceed to offer an outline of my answer to it.

0.2 The Model-Based Account of Psychiatric Diagnostic Reasoning

The basic idea behind my answer to the Methodological Question is that diagnostic psychiatric reasoning can largely be understood as a modelling process that informs less complicated inferential follow-up processes. Therefore, I call my proposal the *model-based account of psychiatric diagnostic reasoning*. Modelling in this context does not mean modelling in the basic sense in that all cognition may be a form of modelling on some level of description, as a number of psychologists and philosophers have claimed (e.g., Johnson-Laird, 2010 Hohwy, 2013). What I have in mind is that on an explicit personal level, the reasoning and actions of psychiatrists in the context of diagnostics embodies the epistemic activity of modelling as we also see it at work in applied or pure sciences.

A proposal to understand psychiatric diagnostic as a form of modelling is overdue. As Mebius and colleagues have pointed out for philosophy of medicine, diagnostic reasoning as "related to model-based reasoning in science [...] is an underexplored area in philosophy of EBM [Evidenced Based Medicine]" (Mebius, Kennedy, and Howick, 2016, p. 760).[2] Although my proposal is more modest, in that it makes no claims about medical diagnostics in general, it at least addresses this issue for the medical subfield of psychiatry. Let me now provide a first rough first idea of what I will argue for.

According to the model-based account of psychiatric diagnostic reasoning, diagnostic reasoning is a multi-level inferential process: a modelling procedure carried out by psychiatrists based on their psychopathological, common-sense psychological, and medical background knowledge. The lowest but also most inference-heavy level of the process is a modelling procedure. On this level, psychiatrists consider the initial presentation of patients leading them to evaluate those patients in more depth for the presence of specific psychopathological symptoms. For this procedure, psychiatrists employ models of psychopathological conditions as well as alternative explanations for patients' problems, and they compare these models to the patients' reports, behaviours, and sometimes cognitive or biological testing. Selecting the best fit from amongst the sufficiently well-fitting of the available diagnostic models, the comparison allows them to infer the presence or absence of specific psychopathological symptoms in patients.

In a second step, the selection of models applicable to the patient – each one applicable to one of the different complaints of the patient and thus suggesting how an aspect of the patients' presentation should be evaluated – is summed up in a bundle of selected models, each of them suggesting a psychopathological evaluation of one of the patient's complaints, accompanied by information about the relevant evidence that led to their selection This synthesis of diagnostic outcomes and supporting evidence is noted in the diagnostic case formulation, which also provides an interpersonal means for clinicians to discuss and assess diagnostic conclusions and reflect on their own diagnostic conclusions.

At the same time, in well-trained clinicians, a pattern-recognition process occurs based on the outcome of the diagnostic modelling procedure that enables clinicians to recognise the diagnosed patterns of symptoms, in accordance with the rules of a presupposed classification system like the DSM or ICD, as syndromal diagnosis takes place. The specific rules governing this process are thereby determined by the relevant diagnostic manual. What exactly this modelling process looks like, as well as

2 To my knowledge, this option is mentioned only in passing by Upshur and Colak (2003) in a general discussion of medical reasoning, and developed only briefly for psychiatric reasoning in Dominic Murphy's Psychiatry in the Scientific Image (2006, pp. 205–209, 365–366).

how it is transformed into the summary case formulation and informs inferencing about the disorder diagnosis, will be developed in detail in the relevant chapters.

As the name of my proposal suggests, the bulk of my work will focus on the level of model-based symptom diagnostics. There are three reasons for this. First, this portion of the diagnostic process is the one where most of the heavy lifting is done in terms of information-gathering and inferential work. The higher levels of diagnostic reasoning – that is, providing the formulation and identifying the right disorder diagnosis based on the rules of a diagnostic manual – are comparatively less complex and will therefore take up less space in my inquiry. Second, the focus on the modelling portion seems appropriate since this most basic level of diagnostics, which provides diagnostic conclusions about symptoms, is the foundation for all higher-order judgements about the presence of psychiatric syndromes. As such, diagnostic modelling will be the source process enabling both formats of diagnostic conclusions: those directly concerned with symptoms and also those indirectly concerned with disorders (i.e., with established conclusions about disorders based on present symptoms). Modelling is, in this sense, the foundational level of diagnostic reasoning. Third and finally, this focus is of interest since the aspect of psychiatric diagnostics that modelling will explain in my account, namely the diagnose of symptoms, has been long neglected in philosophy of psychiatry. The major debates that have raged over the last decade in science, philosophy, and the media since the launch of the DSM 5 have mostly been caught up in debates about disorders. A stronger focus on symptoms will offer a valuable corrective counterpoint to this bias.

Unsurprisingly, my proposed model-based account of psychiatric diagnostic reasoning is not the only game in town, so whatever proposal I make I will have to engage with the other proposals out there. This will be done in full in the final chapter of this thesis. However, to offer an outline of who will be part of the conversation, I will briefly introduce the work of authors who have defended their own positions in response to (aspects) of the Methodological Question, or who have at least been interpreted as aiming to do so.

The first philosopher whose work I will discuss, since it has been taken to contribute to the Methodological Question, is Cooper (2014). Her contribution focuses on case histories in the context of clinical diagnostic work, and more particularly on the role that "Einfühlung" or empathy in understanding one's patient plays in such case histories and how it allows clinicians to provide explanations for patients' clinical presentation. Secondly, we turn to Murphy (2006), who defends a theory of diagnostic reasoning based on the assumption that psychiatrists have fully fleshed-out scientific models of psychiatric disorders from which they derive a further idealised theoretical representation of this disorder and compare this representation of the disorder to the patient to make diagnostic inferences. Then there is Reznek (1998), who decades ago was already seriously engaging with the question of how exactly psychiatrists' diagnostic judgements come about and are justified. Reznek puts forward a proposal that considers psychiatric diagnostics to employ a certain pattern of inference: inference to the best explanation. Next up are Gupta, Potter, and Goyer (2019), who, rather than providing a full account of psychiatric diagnostic reasoning, defend the claim that the second-person perspective, and with it second-person knowledge about the patient that is acquired by empathising with them, is a necessary component of any psychiatric assessment of the presence of mental (i.e., not merely behavioural) symptoms of mental illness. They argue that this is an important enabler of diagnostic reasoning, often missed by existing accounts. Finally, I will consider researchers working within the phenomenological tradition, namely Fuchs (2010) and Parnas, Sass, and Zahavi (2013). They defend an account of psychiatric diagnostics inspired by gestalt psychology in which psychiatrists arrive at diagnostic conclusions by recognising the gestalt of the disorder in the patient's presentation and by this means infer the presence of this disorder.

In Chapter 5, I will argue that the *model-based account of psychiatric diagnostic reasoning* offers a preferable alternative. Now that my own proposal has been sketched out and the discourse about the Methodological Question that it will contribute to has been outlined, let me set out how I intend to structure the presentation of my argument.

0.3 Book Chapter Overview

In Chapter 1, I provide a descriptive account of the core aspects of proper, contemporary, psychiatric diagnostic reasoning and how they are functionally linked to each other in diagnostic practice. This will provide my inquiry with the idea of the diagnostic practice targeted by the Methodological Question and thus by any proposal aimed at answering it. To ensure that the description accurately covers what currently is considered to be proper diagnostic reasoning, this presentation will, as noted earlier, be based on psychiatric training literature as well as the diagnostic manuals and guidelines generated by expert organisations.

In Chapter 2, I introduce modelling in general, and more specifically the form of modelling that I will claim to be the method at work in psychiatric diagnostic reasoning, namely qualitative, constitutive diagnostic modelling. Moreover, I provide an analysis of the rationale behind this method of modelling, and suggestions as to how its conclusions are deemed justified. By addressing these issues, I am able to demonstrate that the method of diagnostic modelling presented in this chapter does indeed map onto psychiatric diagnostics, as well as putting in place the other elements required to provide a full answer to all three aspects of the Methodological Question. That is to say, I will have presented a description, a rationale, and a justificatory analysis for the method of modelling that I need to map onto psychiatric diagnostics.

In Chapter 3, I establish the mapping between the method of diagnostic modelling presented in Chapter 3 and the understanding of diagnostic reasoning laid out in Chapter 1. This mapping supports my proposal that psychiatric diagnostic reasoning should be understood as an instantiation of a specific kind of diagnostic modelling, and that its methodology can be understood along the lines also presented in this chapter. This establishes my initial argument for the plausibility of the *model-based account of psychiatric diagnostic reasoning*.

In Chapter 4, I show how my proposal is able to address the aforementioned desiderata for an answer to Methodological Question. Specifically, I demonstrate that my proposal shows a close match between particular aspects of the method and the diagnostic procedure, leaving no aspects of the diagnostic procedure unexplained; that it provides a cognitively realistic proposal; that it allows for differentiation between a working diagnosis and diagnostic malpractice; that it explains the occurrence and resolution of diagnostic uncertainty; that it allows us to understand and evaluate diagnostic "instincts"; that it explains diagnostic disagreements and their resolution; and finally, that it enables discussion of the impact that progress in psychopathology might have on psychiatric diagnostics.

Finally, in Chapter 5, I look more closely at the alternative proposals touched on earlier that are considered to provide answers to the Methodological Question, some taking similar angles to my approach. I present these accounts in more depth and, for each, indicate specific respects in which the *model-based account of psychiatric diagnostic reasoning* can be considered an improvement on it. The improvement may derive from the fact that an alternative does not actually address diagnostic reasoning (Cooper), makes some implausible moves or is highly abstract (Reznek, Murphy), or relies on claims about parts or the whole of psychiatric diagnostic reasoning that can be shown to be implausible (Gupta, Potter, and Goyer; Fuchs; Parnas, Sass, and Zahavi).

I close my inquiry in the concluding Chapter 6 with a brief review of my argument. I ask whether my research has established an answer to the diagnostic question that meets the criteria set out in this Introduction and whether it offers an attractive alternative to existing views on the details or the entirety of psychiatric diagnostic reasoning.

0.4 Conclusion

In this Introduction I have introduced, motivated, and explained the research question of my investigation. The Methodological Question can be formulated as follows: "What is the method of proper, contemporary, psychiatric diagnostic reasoning?" I have briefly presented the answer to the Methodological Question that I will develop and defend throughout this thesis, as well as offered an outline of other approaches participating in the debate this work contributes to. Finally, I have laid out the roadmap of the thesis, indicating the job that each chapter is doing as part of the whole. I hope that all this will have provided a good framing for the relevance and context of this project and the general direction it is taking.

1. Psychiatric Diagnostics

In this section I will present the core procedures of clinical psychiatric diagnostics. To provide this overview at the outset of my inquiry serves the purpose of gaining a picture of the epistemic practices whose methods I have to account for in order to answer the Methodological Question. To find out what method is at work in psychiatric diagnostic reasoning, getting an idea of how it works as a basis for my inquiry, seems a natural way to begin.

To structure the presentation of psychiatric diagnostics, I will start from the standard boxology model for the general medical diagnostics that scientist and philosophers alike have long supported as the basic framework for thinking about the diagnostic process (e.g., Feinstein, 1964; Elstein, Shulman, and Sprafka, 1978; Sober, 1979). This model carves up the diagnostic process into a three-step input–processing–output format consisting of diagnostic information-gathering, diagnostic information-processing, and, finally, the output of a diagnostic proposal (Figure 1).

Figure 1: The steps of the diagnostic process from beginning to end. Order of progressing steps indicated by arrows.

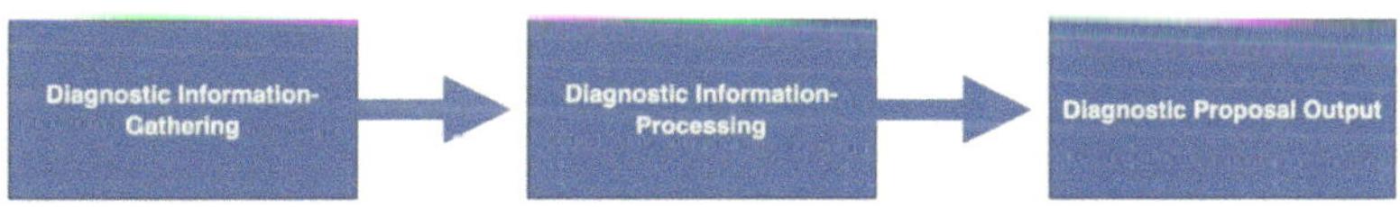

In my presentation I will work though this chart and "unpack" each of these boxes for the case of psychiatry in more detail. For this I will first (1.1) focus on the core practices of diagnostic information-gathering, then (1.2) discuss the diagnostic proposal, before (1.3) discussing diagnostic information-processing, before (1.4) I make a link to the next chapter by introducing the topic of modelling.

Note that the order of my discussion differs from that presented in the flow chart in Figure 1. While it is possible to present the central procedures of diagnostic information-gathering and the format of results from diagnostic efforts, what happens during diagnostic information-processing is more elusive and often only vaguely discussed in the literature. To arrive at an informative picture of this step, inferences based on the more easily explicated input and output steps will be important. Therefore, I will spell out instances of diagnostic information-processing last.

1.1 Diagnostic Information-Gathering

To present the process of diagnostic information-gathering I will concentrate on practices commonly required to be employed in a comprehensive psychiatric assessment. These commonly required components are the *mental status examination*, the *psychiatric interview*, and *cognitive and biological testing*.[1]

Figure 2: Core practices of diagnostic information-gathering.

The mental status examination (MSE) and the psychiatric interview are both necessary components of a comprehensive psychiatric assessment, which in combination are often sufficient to gather the diagnostic information necessary to support a psychiatric diagnosis. In some cases, however, additional cognitive and/or biological tests will be considered necessary to include in a corpus of diagnostic information permitting diagnostic conclusion. As implied, none of these three components alone is considered sufficient to gather the information to provide a diagnosis; a combination is always needed.[2]

1 Note that every psychiatric patient also goes through an initial physical examination, which I do not discuss here since it is not specifically a part of psychiatric diagnostics, rather something that is done with any patient who seeks specialist medical treatment. The purpose of this examination is to prevent nonpsychiatric medical problems from going untreated because they do not surface in patients' complaints, and/or to prevent physical complaints from being wrongly attributed to mental disorder (for example, a complaint about pain might wrongly be considered to be part of a psychosomatic disorder). For more on this latter problem, called "diagnostic overshadowing", see Garden, 2005; Jones, Howard, and Thornicroft, 2008.

2 There are hopes that in wake of "the third wave of biological psychiatry" (Walter, 2013), new methods – such as in genetics and neuroimaging (e.g., Kapur, Phillips, and Insel, 2012) – might soon allow for stand-alone biological tests to diagnose mental disorders. Currently,

Let us look at the basic intention behind each core practice and their implementation in more detail. On the one hand, we have two methods of diagnostic information-gathering that are carried out in a face-to-face examination of the patient: the MSE and the psychiatric interview. The purpose of the MSE is to evaluate the different domains of cognitive functioning such as perception, memory, thinking, affect, time orientation, and thought order, looking for psychopathologically relevant anomalies. This is done by the psychiatrist by observing the patient's behaviour as well as listening to the patient's self-reports in response to specific questions (Trzepacz and Baker, 1993; Casey and Kelly, 2019). The general idea behind the psychiatric interview, by contrast, is that the psychiatrist seeks a broader scope of self-report-based information about the current and past psychological and social functioning of the patient, including factors such as their employment situation, friends and relationships, housing situation, forensic history, substance abuse, sex drive, eating behaviour, and sleeping habits, as well as more systematic background information, for instance about the patient's family history, education, and previous medical problems (Poole and Higgo, 2017; Boland, Verdiun, and Ruiz, 2021).

On the other hand, we have cognitive and biological testing. The first is employed by the psychiatrist to evaluate the cognitive performance of patients in a standardised manner; the second employs biological measures to evaluate the presence or absence of markers that suggest the presence or absence of disorders. The cognitive testing is done by structured examinations consisting of questions to be answered by the patients (e.g., "what day is it today?") and cognitive-behavioural tasks to be executed (e.g., "please remember and repeat the following words"; "pick up the pen with your right hand and draw this clock") whose outcomes are scored and compared to cut-off criteria to decide whether anomalies are present. The *Cambridge Cognitive Examination Revised* (Roth et al., 1998) and the *Alzheimer Disease Assessment Scale* (ADAS) (Hodges, 2017) are examples of such tests relevant to supporting the diagnosis of dementia or other neurodegenerative disorders.[3] Biological testing, meanwhile, uses specific biological markers to indicate the presence or absence of specific conditions

however, psychiatry has not yet established biomarkers for clinical use allowing us to arrive at unambiguous diagnostic conclusions about the presence of disorders, let alone of specific symptoms (Martins-de-Souza, 2013; First et al., 2018; García-Gutiérrez et al., 2020).

3 By talking about neurodegenerative disease as psychiatric disorders I do not want to take a stance in the ongoing ontological debate whether mental disorders are brain disorders (e.g., Boorse, 1977; Papineau, 1994; Insel and Quirion, 2005; Miller, 2010; Graham, 2013; Schramme, 2013; Insel and Cuthbert, 2015; Olbert and Gala, 2015; Jefferson, 2020), or the related debate in the medical community as to whether we should distinguish between neuropsychiatric or psychiatric disorders in the clinical context (e.g., Price, Adams, Coyle, 2000; Baker, Kale, Menken, 2002; David and Nicholson, 2015). Instead, I simply adopt the current standard of psychiatry itself, whose current boundaries encompass neurodegenerative disorders, making these diseases part of the current responsibility of psychiatry.

that can inform (differential) diagnostics of disorders responsible for behavioural and mental alterations in patients. Relevant for this are serological testing, genetic testing, and radiological examinations. To offer a few examples: Liquor analysis can reveal levels of β-amyloid, total tau, and phospho-tau-181 that indicate the presence of irreversible forms of dementia (Reitz and Mayeux, 2014). Genetic testing can show whether patients are carriers of ultra-high-risk genes for developing Huntington's disease (Myers, 2004). Neuroimaging data can be important in identifying strokes or major structural alterations of brain substance that may be responsible for cognitive and behavioural alterations (Power et al., 2016; First et al., 2018). Neuroimaging data also allow us to distinguish between the subtypes of prefrontal lobe dementia versus Alzheimer's (Rohren et al., 2013).[4] Again, such testing mainly supports the diagnosis of neurodegenerative disease, but it can, in addition, be especially relevant to enabling differential diagnostic conclusions that reveal a psycho-behavioural condition to be a nonpsychiatric case – for example, if the patient is found to have a brain tumour that can be assumed to cause their condition.

Figure 3: Core practices of diagnostic information-gathering mapped onto their categorisation as contributing to diagnostic screening and in-depth evaluation.

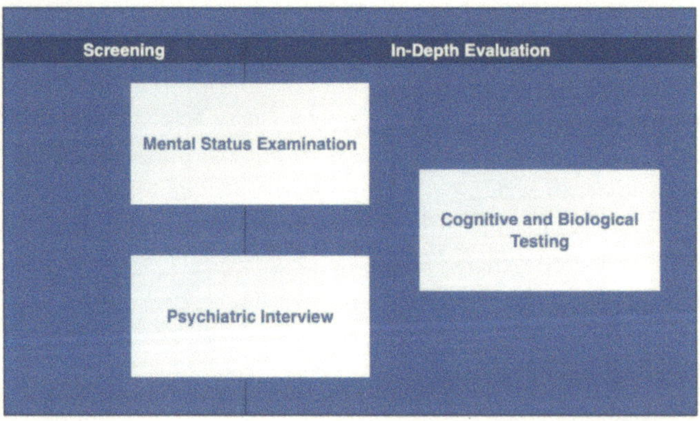

4 I am aware that success in this domain of diagnostics is still limited insofar as this method does not yet yield good results in differentiating between Alzheimer's disease and forms of dementia other than the prefrontal type, such as Lewy body, frontotemporal, and vascular dementias (Maclin, Wang, Xiao 2019). This innovation is also an outlier in the field of research on neurodegenerative disease, where so far nothing similar has been achieved for Parkinson's (Miller and O'Callaghan, 2015; He et al., 2018), Huntington's (Silajdžić and Björkqvist, 2018), or amyotrophic lateral sclerosis (Verber and Shaw, 2020).

In addition to breaking down the information-gathering procedures of psychiatric diagnostics into its component parts, we can further specify the process they are employed in as consisting of two functionally different stages cutting across these components: the *screening* and the *in-depth evaluation*. While the MSE and the psychiatric interview contribute to both screening and in-depth evaluation, cognitive and biological testing is solely a method of in-depth evaluation.

The purpose of screening is to arrive at a list of the patient's *complaints*, which can subsequently, via a more in-depth evaluation, be judged to be psychiatric symptoms/signs or not. By "complaints" we should not only understand things that the patients themselves complain about; these would be *subjective* complaints. The category of complaints also encompasses the *objective* type – that is, psycho-behavioural obstructions that are recognised by the psychiatrist but may go unrecognised by the patient. The list of complaints is formed by paying attention to *prima facie* obstructed aspects of the patient's psychology and behaviour that in light of psychopathological background knowledge appear to be similar enough to psychopathological phenomena to justify a more careful examination to determine whether they are indeed psychopathologically relevant symptoms and signs. As this suggests, complaints in themselves are not automatically considered psychopathologically relevant signs and symptoms; they are mental or behavioural features of the patient noted by the psychiatrist as deserving a more in-depth evaluation in the context of the assessment. This in-depth evaluation is then conducted in the same face-to-face setting and possibly supported by additional cognitive and biological tests. In this in-depth assessment, further information allows the psychiatrist to decide whether the complaints under consideration should be assessed as psychopathological symptoms/signs; psychological or behavioural problems resulting from medical non-psychopathological problems; or psychological or behavioural complications of no medical relevance at all.[5]

Let us look more closely at what a screening procedure followed by in-depth evaluation will usually look like. The first thing to point out is that in clinical practice the

5 Why should two hypothetically similar instances, behaviours, or mental states be classified as a psychiatric symptom/sign on one occasion and a non-psychiatric one on another? There is no strong metaphysical reason, but in the special place that psychiatric symptoms and signs currently have in medical semiology (Altable, 2012). In medicine, symptoms are traditionally considered manifestations of a disease, or to put it more philosophically, they are representations of the presence of these diseases, and therefore of physiological alterations considered causally responsible for their presence. If a symptom or sign is caused by a disease condition that is not considered a mental disorder, then for the clinical purpose of providing diagnosis of psychiatric disorders it is not considered to be psychiatric sign or symptom. This does not mean that research might not ultimately show that part of the causal pathways responsible for the occurrence of the symptoms is shared by a psychiatric disorder and a disease with similar psychological or behavioural symptoms.

MSE psychiatric interview (as the two components of screening) are often conducted in the same face-to-face encounter, as are their in-depth evaluation elements, while the in-depth evaluation by means of cognitive and biological testing is often conducted on another occasion. This order of things has pragmatic reasons that, though not imperative,[6] come to bear often enough to consider a default. The pragmatic reasons for this are that much of the initial screening information considered relevant to the domains covered by the MSE can also be covered within the face to-face interview situation of a psychiatric interview, so that it is economical to conduct them together. By contrast, conducting biological tests or preparing and administering cognitive tests takes time, so that a special appointment is usually needed.

To get a better grasp on this combination of the MSE and psychiatric interview assessment, let us consider an instance in which both are combined. The assessment begins in the moment when the patient and the psychiatrist meet. From the first moment onwards, the psychiatrist observes the patient in light of his/her psychopathological background knowledge and clinical experiences, seeking a first impression of the patient's psycho-behavioural setup in order to recognise conditions that *prima facie* may be potentially psychopathological relevant. The focus hereby lies on aspects of the patient relevant for the MSE: body posture, facial expressions, movements, and gaze behaviour are some of the earliest parameters relevant to recognise in order to glean an idea of things like the patient's mood, psychomotor-activity, and wakefulness.

As the conversation begins, the psychiatrist will typically open the interview with an open question like, "what is the reason for your visit?", to invite the patient to report on what brought them to psychiatric services. The content of the answer to this question will then be the main source of information about subjective complaints that may turn out to be symptoms. If this initial question is answered, the psychiatrist usually addresses further domains of psychological and social functioning to make sure that there are no complaints that might not have been mentioned so far by the patient, which may be the case if patients themselves do not considered complaints to be relevant or have forgotten to mention them. Some people, for example,

6 Note, however, that there is some variety in style and preference among clinicians. Some preferer to first conduct a full MSE and then a full Interview, while others combine them. Some like to do the screening and in-depth evaluation in one encounter; others like to or have to split the evaluation into multiple sessions because of time limits or because the patient has difficulty focusing on the process. Some like to first get a full overview of present complains in patients and then come back to each to each noted complaint for an in-depth evaluation; others like to interrupt the screening if a complaint is noted and go into greater depth right away. I take these differences in style to be accidental differences leave untouched the essential distinction between the functions of screening and in-depth evaluation that are served by different aspects of the assessment, however one may prefer to carry it out.

do not consider it noteworthy that they have sleeping problems because they have had such problems on and off their whole life.

While the patient is questioned by the psychiatrist, observation carries on, now also picking up potential abnormalities in the form and content of the patient's language production. Here the psychiatrist may recognise various forms of linguistic alterations that fall within the scope of phenomena whose recognition is part of the MSE's purpose. For example, the psychiatrist may recognise increased rates of speech: an objective complaint that can turn out to have symptom value as "pressured speech".

Finally, once all screening questions relevant to the psychiatric interview have been asked, the MSE-relevant observations that have been made will be complemented by the psychiatrist asking questions and giving tasks to the patient to cover remaining aspects of the MSE that so far have not been dealt with. This is often done at the end so as not to interrupt the flow of conversation during the interview. Asking and tasking will target specific domains of cognitive or behavioural functioning that could not be observed sufficiently during the interview. Often the psychiatrist will, for example, explicitly screen for semantic memory deficits by tasking the patient to name objects in the room or will evaluate their orientation in time by asking "what day and month is it today?".

Once the screening is done, the psychiatrist, equipped with a list of the patient's subjective and objective complaints, will turn to the in-depth evaluation, as far as it can be carried out in a conversational setting. In the in-depth evaluation, noted complaints will be targeted in more detail, based on the psychiatrist's hypothesis as to which symptoms and signs might be present in the patient and which alternative non-psychopathological state of affairs might have led to their occurrence given the psychiatrist's background knowledge (a form of differential diagnostics). If cognitive and biological tests are thought to be relevant, they will also be conducted with the patient. Within the face-to-face evaluation, the psychiatrist will be interested in generating a more detailed description of self-reported experiences and behaviours that lead to the initial assumption of the complaints. This will include information such as how long the complaints have been present, or when they appear and whether they are always the same or change under certain circumstances. The psychiatrist will also try to attain information that the patient themselves might not connect to their condition – for example, the presence or absence of typical aetiological factors, or a typical consequence of a psychopathological condition that would match with the present complaint. Information from potentially conducted cognitive and biological testing, such as test scores from formal memory assessments or neuroimaging or serological data that might inform inferences about brain lesions or non-psychopathological causes of psycho-behavioural alterations, will be waited for and taken into account. These complementary forms of evidence allow the psychiatrist to draw conclusions about which of the complaints should be assigned psychiatric symptom/sign status.

So far, I have presented a general description of the content and purpose of the three core aspects of diagnostic information-gathering, and I have offered a bird's-eye view on how they are conducted in order to establish the distinction between screening and in-depth evaluation. While this may suffice to gain a general idea of this step in psychiatric diagnostics, I will now introduce a set of more detailed showcases for the recognition and evaluation of complaints, for each of the three lines of in-depth evaluation. These more detailed examples will be used later to support my ideas about how to best interpret this step of the diagnostic process in terms of a theory of diagnostic reasoning, a task for which a bird's-eye description alone is too abstract. Please note that in my examples I will also indicate what conclusions the psychiatrist may draw regarding what symptoms and signs are present in patients based on the in-depth evaluation. I do so to provide a more organic picture of the process of diagnostic information-gathering and the role of the in-depth evaluation. Strictly speaking, information-gathering ends with in-depth evaluation, but to break off in the detailed description at that point makes it hard for us to grasp what is really going on. How exactly the psychiatrist moves from the end of the in-depth evaluation to their conclusions regarding present symptoms and signs will be something I will come back to in detail when I discuss diagnostic information-processing and the generation of diagnostic proposals.

Let me begin with an example of the screening and in-depth evaluation that would formally be considered part of the MSE. Imagine that over the course of the interview, the psychiatrist's attention to the patient's language production suggests a formal anomaly. The patient shows a significant deficit in amount of spontaneous speech, manifested in the form of very brief, concrete, and unelaborated answers to questions. The following table offers an example of the evidence that might be taken to suggest this type of anomaly.

Table 1: Example conversation illustrating the difference between the speech pattern of a patient who is likely to be suspected of suffering from a psychiatric complaint ("Anomalous") versus a non-noticeable example ("Normal").

Anomalous speech pattern	**Normal speech pattern**
Psychiatrist: Good Morning, Mr X. What can I do for you?	**Psychiatrist:** Good Morning, Mr. X. What can I do for you?
Patient: You can help me.	**Patient:** I came to you because I have some problems that I think I need help with. (...)

Psychiatrist: And I will try my best to do so. Can you tell me something about the reason why you reached out for help?	**Psychiatrist:** And I will try my best to do so. Can you tell me something about the reason why you reached out for help?
Patient: Yes	**Patient:** Well, thanks. I feel sad and empty, and I don't know what I should do about it. It started (...)

Such unusual verbal response patterns will make the psychiatrist consider the patient from the perspective of a complaint of reluctant speech that on closer examination may turn out to be "poverty of speech", a form of alogia. Alogia is a psychiatric symptom that can involve impoverishment regarding the quantity of speech – *prima facie* matching the presentation of the described case – or regarding content of speech and thought, such that the number of topics the patient is able to cover is seriously limited. Alogia is considered to be present in various mental and neuro-psychiatric disorders such as dementia, schizophrenia, severe depression, or schizotypal personality disorders (APA, 2013, p. 817).

Whether the psychiatrist concludes that the patient indeed suffers from this condition will again depend on a closer evaluation. For example, this sort of behaviour may be evaluated as forming part of her usual premorbid behaviour, as is sometimes the case in people who are unusually pedantic in their speech – a habit that may evoke the impression of poverty of speech. This is a problem that has been observed in administrators, politicians, scientists, and of course philosophers (Andreasen, 2016). If this appears to be the case, the complaint would *prima facie* not qualify to be evaluated as a case of alogia. The same would be the case if the patient felt discomfort or anxiety in the interview situation that seemed to lead her to choose his words carefully and use them sparely. On the other hand, if the psychiatrist finds these two options to be ruled out by a more in-depth evaluation of the patient's emotional attitudes towards the interview situation, as well as her premorbid use of language, the psychiatrist may proceed to conduct a cognitive assessment of the patient's cognitive capacity to produce certain patterns of language use, and perhaps to test for specific cognitive processing capacities whose impairment is associated with alogia. This should allow the psychiatrist to decide whether the conclusion that the patient's complaint indeed is a form of alogia may plausibly be drawn.

The chief cognitive impairment underlying presentations of alogia in psychiatric cases seems to be an impairment of *control retrieval* – an aspect of the executive function allowing the individual to retrieve information from memory when the information is not automatically retrieved and when there is more than one potential unit of information that would match the search profile that could be activated (Wagner et al., 2001; Doughty and Done, 2009; Docherty, Berenbaum, and Kerns, 2011). If a test of speech production carried out with a cooperative patient shows patterns indicating the corresponding kind of cognitive impairment, the conclusion that the patient suffers from alogia seems warranted. This can, however, be tested using *verbal fluency tasks* in which subjects are given a production rule for producing words that, for example, begin with a certain letter (testing word letter fluency) or fall into a category such as animals (testing word fluency). In our case, subjects would be asked to produce items for a certain span of time. A deficit in this task shown by individuals who suffer from cognitive impairment of control retrieval is an increased mean response latency between each reported word if asked to produce words in a given category, but in absence of deficits suggesting the impairment of other language-related cognitive functions that in principle could also lead to the clinical presentation. These might include disorganised semantic memory, which would lead to poorer performance on category fluency relative to letter fluency, and context processing, which leads to a decrease in the proportion of correctly reported semantically related words (Docherty, Berenbaum, and Kerns, 2011). If verbal fluency testing of the patients meets this prediction, a conclusion that the patient's complaint is an instance of the symptom of alogia may be drawn.

Next, let me consider an example that might come up in the context of the psychiatric interview. Consider a patient reporting sleeping problems, either in response to the opening question, or following superficial checking of domains of psychological and social functioning in which context the psychiatrist will also ask whether the patient sleeps well. Psychiatrists ask this question because sleeping problems are of diagnostic importance, on the one hand since they occur in the context of various mental disorders such as depression, PTSD, and anxiety, which can be related to different patterns of sleeping behaviour (Krystal, 2012), but also because specific types of sleep disturbance can even be relevant to subtypes of major mental disorders. For example, hypersomnia is associated with atypical depression and terminal insomnia is related to melancholic depression (Murphy and Peterson, 2015), making it important to have a proper grasp of a patient's sleep-related symptoms in the interests of accurate differential diagnostics.

To determine whether a patient's complaint of sleeping problems qualifies as a psychiatric symptom requires a detailed evaluation, however. The psychiatrist will ask about specific features of sleeping behaviour, such as whether the problem is with falling asleep, getting up, or sleeping through the night and whether this leads to unusually short or long periods of sleep or an atypical sleep rhythm, as well as how long the patient has had these problems and whether they occur only occasionally or on a regular basis. To find out how this problem might relate to other behaviours and experiences, the psychiatrist will ask how the patient feels before he goes to bed, and whether there is something the patient does only on the occasions when he does not sleep well. Based on this information, the psychiatrist will then decide whether the complaint should be considered a psychopathological symptom or non-psychopathologically relevant, or maybe even a non-psychiatric medical problem. If, for example, the patient reports drinking half a gallon of cola and watching Netflix in bed before they try to sleep, in the evaluation their complaint will *prima facie* not be considered a psychiatric one, and if the patient ends up reporting that their problems with falling asleep began around the time when they started to take beta-blockers to treat their high blood pressure, again the psychiatrist will consider the sleeping problems a drug side-effect rather than a psychopathological relevant symptom. If, however, none of these scenarios applies, but instead the patient reports increasing agitation and worry in the evening hours that cannot be stopped intentionally, leading him to feel unable to sleep so that he begins to drink to calm himself down and then finally be able to fall asleep, the psychiatrist will tend to judge the complaint to qualify as a psychiatric symptom, due to its apparent relation to other cognitive and behavioural complaints prototypically associated with psychopathological cases of insomnia (see e.g. Krystal 2012).

To consider a case in which cognitive or biological testing makes a significant contribution to the outcome of an in-depth evaluation, let us look at a patient who has reported often feeling very tense and who is experiencing anger and has outbursts of aggression in response to minor stressors, such as not finding her keys or being asked to repeat something because she spoke too quietly. This initial description of the complaint encourages the idea that the patient might present psychiatric symptoms/signs of *irritability*, which is diagnostically relevant for 15 disorders of the DSM-5, including mood disorders, addictive disorders, and personality disorders (APA, 2013). Irritability itself may be understood as a "partial physiological agitation characterized by an increased sensitivity to sensory stimuli and a non-cognitively mediated lowered threshold for responding with anger and/or aggression to typically less vexing stimuli […]" (Toohey and DiGiuseppe, 2017, p. 31). Sometimes psychological research considers irritability as a *state* of mind (e.g., *irritable mood*; Toohey and DiGiuseppe, 2017), whereas on other occasions, for example in developmental pathological research, it is mostly discussed as a *trait*, e.g. *irritable personality* (Leibenluft and Stoddard, 2013).

However, not all instances of irritability appear to be clinically relevant or to qualify as a psychiatric symptom. Indeed, irritability itself is a well-known psychological phenomenon. All of us will at some point have felt tense because we were hungry, in pain, or exhausted, and we have probably all lashed out, in that state, at someone who did nothing particularly wrong, but no psychiatrist would be keen to attribute the symptom of irritability to us based on such instances. Rather, it seems that from a clinical perspective, the symptom value of irritability has to be excessive in its rate of occurrence and the degree to which it interferes with psychosocial functioning and impairs the individual's capacity to effectively and quickly handle tasks. A clinically irritable person will also typically be expected to show increased biases to attribute hostile and bad intentions to other neutral or even friendly individuals, as well as a tendency to develop anticipatory frustration for future events, often leading to feedback loops increasing the level of negative expectations (Yager, 2020). To see whether this matches the current case, let us come back to our example patient.

The psychiatrist will ask the patient when the irritability first appeared, how often it occurs, and how it influences the patient's daily life and her interactions with others. The psychiatrist will also ask about the patients' social relationships and how she is doing at work (if these areas have not already been covered) and will try to find out whether the patient shows patterns of negative attributes that are hard to explain with reference to particular experiences the patient is able to cite. Imagine that the psychiatrist hears from the patient that the irritability surfaces every other day and persists for a few hours, thereby seriously impeding progress in work tasks and also making it hard for her to deal with her coworkers or be at home with her young children. However, the patient does not seem to be very negative in her orientation to others, but rather thinks that the problem is in herself. Often the irritability is accompanied by sweating, and sometimes by blurred vision, and there is no evidence that the patient has any obvious other condition such as problematic eating patterns or chronic pain that could account for the irritable mood.

While some parts fit the previously introduced psychiatric clinical understanding of irritability, others obviously do not, so the overall picture appears inconclusive. However, the report of the phenomenon of blurred vision fits with another potential explanation for irritability the psychiatrist is aware of: Irritability can also be a sign of badly managed diabetes, which would also fit with the sweating reaction and usually does not lead to more wide-reaching psychological changes regarding others; it also does not require abnormal eating patterns to arise on a regular basis. Torn between the option of assuming that the patient irritability does not qualify as any symptom (neither a psychiatric symptom, nor a symptom of a non-psychiatric medical problem) and the option that the patient's irritability is symptomatic not as psychopathological symptom, but could rather be the psychological consequence of processes caused by irregularities in her blood sugar levels, the psychiatrist orders tests for diabetes. In case of a positive result, the psychiatrist would not consider the patient's irritability a psychiatric symptom that he would make reference to if he were to match the patient's overall psychopathological condition with DSM symptom requirements. If the test were negative, the psychiatrist would have to consider the question undecided and would be able to conclude only that there is a possible presence of irritability as a psychiatric symptom. After all, there might still be other conditions in the patient whose evaluation may lead to the conclusion that something other than diabetes caused the irritability. Or, indeed, irritability may not be possible to ascribe with certainty, and the patient may suffer from other psychiatric symptoms or signs that might be confirmed after further evaluation. With this remark I close my presentation of cases exemplifying the screening and in-depth evaluation for all three discussed procedures of diagnostic information-gathering.

Having completed the presentation of the information-gathering procedure, two more things are left to do before I turn to unpacking the diagnostic proposal output box. First, I shall briefly respond to an immediate worry that clinicians and scientists may have regarding the adequacy of my presentation,. Then, to keep track of the outcomes of my presentation, I will present an updated version of the initial flowchart (Figure 1) integrating what unpacking the box of diagnostic information-gathering has revealed.

Let me begin with the worry one may have. While I consider that most clinicians will agree that the means of diagnostic information-gathering I have discussed are central to psychiatric diagnostics, some might wonder why other methods, especially questionnaires and structured diagnostic interviews, have been neglected. I have not discussed these methods here because they are not among the *constitutive* core practices of psychiatric diagnostics, but are only of secondary relevance compared to those core practices. By this I mean that employing them (opposed to the core diagnostic methods I have discussed) is not necessary for comprehensive psychiatric assessment, nor are they sufficient to gather the diagnostic information required for a comprehensive diagnostic process.[7] Rather than being part of core diagnostic practices, questionnaires and structured diagnostic interviews are useful complements to them. As questionnaires or structured interviews are complements, including them in my presentation would be redundant, given the explicitly stated aim to focus solely on core procedures of psychiatric diagnostics.[8]

---

7 My understanding of *constitution conditions* thereby draws on the work of Tyler Burge, who argues that pursuing and explicating a phenomenon concerns its constituents if it focuses on the necessary and/or sufficient conditions for something to be what it is: "A constitutive question concerns conditions on something's being what it is, in the most basic way. Something cannot fail to be what it is, in this way, and be that something. Constitutive conditions are necessary or sufficient conditions for something's being what it is in this basic way. To be constitutive, the conditions must be capable of grounding ideal explanations of something's nature, or basic way of being" (Burge, 2010, p. xv).

8 If this claim strikes you as strange or unintuitive, this footnote is for you. To avoid misunderstanding: I do not claim that questionnaires or structured interviews are useless or irrelevant. Questionnaires such as the Beck Depression Inventory (Beck et al., 1961) can be useful for screening, and structured interviews such the Structured Clinical Interview for the DSM-5 (SCID) (First et al., 2016) can help acquire much important diagnostic information. All I want to say is that by looking at in a structured way at psychiatric practice we note that diagnostic information-gathering by questionnaires and structured interviews plays a subordinate in clinical diagnostics, something that is done in the context of psychiatric diagnostics but does not individuate it. Think of questionnaires. A responsible clinician would not make a categorical disorder diagnosis based solely on the answers to a questionnaire, nor can a diagnostic case formulation be provided based on them. However, drawing diagnostic conclusions with-

Now to the modifications of the flowchart. The process of psychiatric diagnostics described above is complex enough to warrant a graphical illustration. Let me briefly recap which aspects of the process the flowchart must do justice to. As I indicated through my presentation, the "diagnostic information-gathering" box contains three procedures: the MSE, the psychiatric interview, and cognitive and biological testing. These three procedures serve two functional distinct roles: screening and in-depth evaluation. The first aspect of screening provides information about the present complaint. In both identifying these complaints and determining how to carry out the in-depth evaluation, the psychiatrist's psychopathological and biomedical background knowledge plays an important role. It therefore seems that some diagnostic information-processing is already taking place between the screening and the in depth-evaluation – an additional stage of "diagnostic information-processing" in the midst of diagnostic information-gathering that did not appear in the initial flow-chart (Figure 1). I therefore propose the following flowchart (Figure 4), as graphical presentation of the overall process I have described in this section and summed up in this paragraph. Next, I will turn

out using questionnaires is not an improper diagnostic practice in psychiatry. Questionnaires can contribute to a diagnosis, but only interpreted in the context of an overall clinical impression, generated from what I consider to be the three core procedures. Now think of structured interviews. Even at first glance, it is clear that they are not a necessary component of psychiatric diagnostics. We rarely find them used outside of research contexts, such that rather than being essential to proper clinical diagnostics in general, they are an essential tool to clinical research (Aboraya, 2009; Bruchmüller et al., 2011; Mueller and Segal, 2014). That they are also not sufficient to make a diagnosis can be shown in two ways. First, structured interviews do not provide the information necessary to provide a case formulation (discussed in the next section) that has to be provided as part of the diagnostic proposal; this needs, amongst other things, more biographical, psychosocial, and other data from patients that is not attained in typical structured interviews but is provided by the psychiatric interview. Second, structured interviews usually ask questions explicitly mentioning symptoms relevant to categorical diagnosis and thereby hope to elicit answers that collectively allow one to make a diagnosis. However, research suggests that experienced clinicians – when they do use such interviews in evaluating patients – take into account not only the answers to these questions, but also a wide range of patient behaviours they observe in their contact with the patient that would usually fall under information collected in the mental status examination (Nakash and Alegría, 2013). The fact that taking into account additional information such as observable behaviour that goes beyond the mere answers to a structured interview in order to establish a diagnosis is not a mere quirk on the part of clinicians but an important aspect of diagnostic practice can be shown by considering what happens if individuals who are not clinical experts use such interviews. Research suggests, for example, that SCID interviews carried out by laypeople who do not have the skill to implement aspects of the MSE interviews have low validity (as measured against the diagnostic judgements of expert clinicians) (Nordgaard et al., 2012). In conclusion, using a structured interview cannot replace the psychiatric interview, nor does it make an MSE superfluous.

to unpacking the "diagnostic proposal output" box at the bottom of the flowchart's current extent.

Figure 4: Modified flowchart of stepwise psychiatric assessment as developed in this section (1.1). Vertical arrows connect steps in the process; horizontal arrows indicate influencing factors.

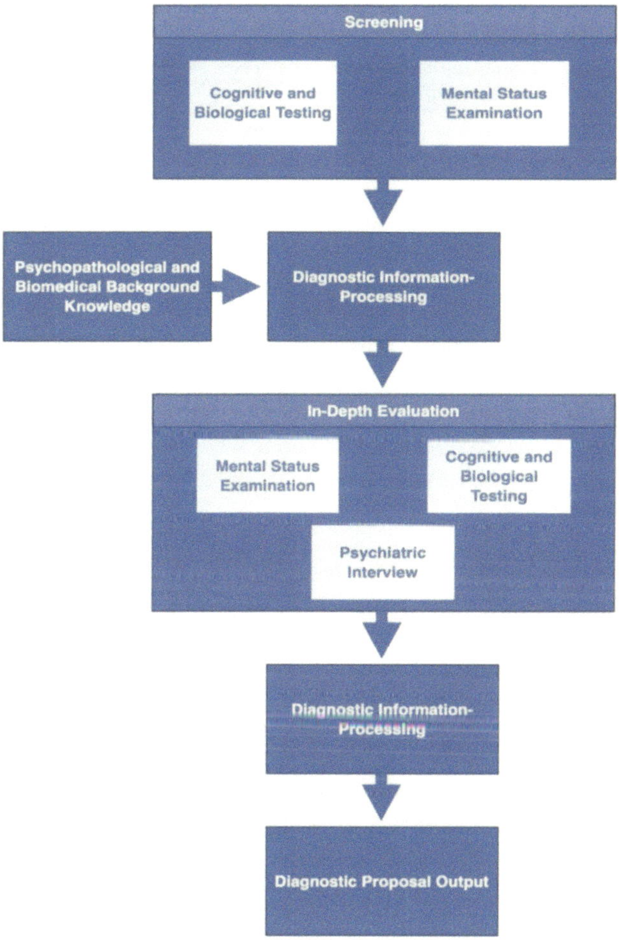

1.2 Diagnostic Proposal (*Output*)

Based on the body of information that psychiatrists collect, they are meant to formulate a diagnostic proposal. Following the American Psychiatric Association (APA, 2013), it should consist of a diagnostic case formulation[9] and a manual-based diagnostic classification of the disorder (Figure 5).

Figure 5: The two components of psychiatric diagnosis.

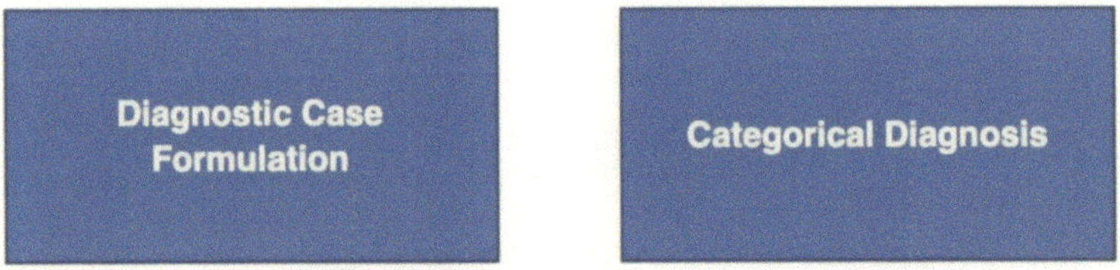

By organising diagnostic information in this way and relating it to the patient's complaints, the case formulation intends to provide a structured presentation of diagnostic information that stands in an explanatory relationship[10] to the patient's complaint, allowing the psychiatrist to determine which aspects of the patient's presentation should be interpreted as presenting which psychiatric symptoms/signs or non-psychiatric problems. As such, the formulation also serves as justificatory grounds for the attribution (or not) of psychopathologically relevant features to the patient.[11]

---

9 Note however that not only does the APA consider case formulation (outside the United States sometimes called clinical formulation) to be a proper part of psychiatric diagnostics, case formulations are widely recognised as a diagnostic standard in psychiatric diagnostics. Official statements and educational guidelines of various expert societies show that they consider it a core competence in diagnostics, and part of good psychiatric practice. See, for example, Royal College of Psychiatrists, 2013, 2017; Royal Australian and New Zealand College of Psychiatrists, 2014; American Board of Psychiatry & Neurology, 2019.

10 What kind of explanation the case formulation is intended to provide and how it is thought to do explanatory work is usually not specified in the clinical literature. I will come back to this issue by making a proposal as to how to understand the explanatory qualities of case formulations in Chapter 5.

11 While approaches differ in terms of what exactly a case formulation should look like, my characterisation here appears representative in its core idea, assumed across the literature on case formulations. To compare, see, e.g., Meyer and Turkat, 1979; Varghese and Mellsop. 1983; Weerasekera, 1996; Butler, 1998; McHugh and Slavney, 1998; McWilliams, 1999; Eells, 2006; Division of Clinical Psychology (British Psychological Society), 2010; Kuruvilla and Kuruvilla, 2010; Johnstone and Dallos, 2013; Bruch and Bond, 2015; Goldman and Greenberg, 2015; Kennerley, Kirk, and Westbrook, 2016; Ryan, 2019.

To gain an impression of this format of diagnostic proposal, let me provide an example from the literature: the case of Mr Z (Sperry, 1992). Here is a brief description of the case of Mr Z, followed by a diagnostic case formulation based on Sperry's discussion of the case.[12]

Case description Mr. Z

Mr Z is a 40-year-old businessman who presented with complaints of loss of interest in his job, hobbies, and family over a period of six weeks. He acknowledged periods of profound sadness, reduced appetite with significant weight loss, insomnia, fatigue, and recurrent thoughts of death, but denied suicidal ideation. He denied any precipitants but did admit that his expected job promotion had not materialised. Mr Z described himself as unusually serious, conservative, and relatively unable to express affection. He also acknowledged trying to be perfect, needing to be in control of every social situation, and having an excessive commitment to work.

Mr Z indicated that his marriage had been worsening for several years and described his wife as flighty, overemotional, and helpless under stress. For the past several years she had been angry, distant, and had declined to be involved sexually with him. Since the onset of his symptomatology, however, she had been solicitous and obviously concerned. The Z's have two children, a boy, 12, and a girl, 10, who appeared to be doing well at school and home.

Mr Z described his family of origin as very poor. His father deserted his mother when the patient was 12 years of age and, as the oldest child, he had to take considerable responsibility for younger siblings, as well as to work part-time while attending school. He knew that his maternal grandfather had committed suicide and that two maternal uncles were alcoholics. A paternal uncle had died in prison after a long period of antisocial behaviour. Physical, laboratory, and neurological studies were negative.

Diagnostic case formulation Mr. Z

Mr Z is a 40-year-old married businessman whose depressive-like symptoms began shortly after being passed over for a promotion. Other stressors appear to be chronic marital and sexual problems and the fact that his two children are nearing the age of independence and the age when he experienced a significant trauma in his own life: the desertion by his father when he was 12.

Although there is a positive family history for alcoholism, suicide, and sociopathy, Mr Z denies other psychiatric symptoms or treatment for himself. Mr Z's family

---

12 The following example is one of the rare high-quality illustrations of the structure of a case formulation, but it may appear outdated in parts to readers familiar with clinical matters. Please take into consideration its age and accordingly the changes in our understanding of psychopathology that have taken place since its publication.

history of alcoholism, suicide, and sociopathy makes it likely that he has a genetic predisposition for affective illness.

He appears to have major conflicts over dependency and autonomy. Because of his earlier experience with significant loss, the withdrawal of attention and affection by Mr Z's wife and the growing independence of his children represent significant precipitating events, Mr Z has considerable difficulty expressing emotions and affection. He is controlling and perfectionistic. His cognitive style is obsessive-compulsive. His primary defences are repression, regression, introjection, isolation of affect, and intellectualisation. Mr Z's sociocultural background has helped to instil in him a basic belief in the value of hard work, stoicism, and self-reliance with little dependence on extra-familial sources of support. From a young age, he has been reinforced to sacrifice himself and to maintain the role of provider and nurturer to others who have depended upon him for support.

Mr Z is also distant from his family of origin and his current life centres on his immediate family. His role has been as a provider to a wife and children who have been dependent upon him. Mr Z and his wife have not been able to form a satisfactory marital coalition, they do few things together, and their sexual relationship has deteriorated. His wife had withdrawn emotionally and sexually from him until his recent problems, which promoted her attention and concern. Mr Z has been able to adapt fairly well educationally and occupationally and is a successful businessman. However, he has limited social relationships, no close friends, and few independent recreational activities.

Mr Z's probable biological predisposition to affective instability, coupled with the abandonment by his father and familial and sociocultural reinforcement, resulted in the development of a rigid, obsessive-compulsive personality. His role evolved into one of stoic, hard-working self-sacrifice in the service of others who are dependent upon him and a denial of his own dependency needs. While adaptive educationally and occupationally, his personality structure and ego defences resulted in an isolated lifestyle and the inability to acknowledge his own feelings or to relate to others with warmth and affection. The symbolic abandonment by his wife and children reawakened old dependency conflicts, threatened his adaptive role in life, overwhelmed his rigid defences, and resulted in anxiety, regression, and depression.

A problem list includes 1) clinical depression; 2) marital discord including sexual difficulties; 3) an obsessive-compulsive style; 4) limited social support system with friends; and 5) limited recreational activities.

The other aspect of the diagnostic proposal is the diagnostic categorisation. The idea here is to categorise a present psychopathological condition based on clusters of signs and symptoms that consist of necessary criteria plus a defined number of further diagnostic criteria from a fixed list of possible items, which in combination are sufficient to diagnose a disorder. The criteria can be either fulfilled (symptom present) or not fulfilled (symptom absent). Every disorder category is mapped onto a set of partly differing combinations of signs and symptoms that have to be present to apply the category to a patient. The criteria to be checked thereby consist of behavioural, emotional, and cognitive features. In some cases, further criteria such as a temporal qualification (e.g., the condition has to be present for at least two weeks) or the requirement of certain types of environmental factor (e.g., experience of a life-threatening, dangerous, or significant abusive circumstance for post-traumatic stress disorder) are explicitly mentioned. To illustrate this aspect of the diagnostic proposal, see the following criteria for major depression disorder from the DSM-5 (APA, 2013, p. 160), which allows for 50 combinations of signs and symptoms to apply this category.

DSM Criteria for Major Depression:
The individual must be experiencing five or more symptoms during the same 2-week period and at least one of the symptoms should be either (1) depressed mood or (2) loss of interest or pleasure.

1. Depressed mood most of the day, nearly every day.
2. Markedly diminished interest or pleasure in all, or almost all, activities most of the day, nearly every day.
3. Significant weight loss when not dieting or weight gain or decrease or increase in appetite nearly every day.
4. A slowing down of thought and a reduction of physical movement (observable by others, not merely subjective feelings of restlessness or being slowed down).
5. Fatigue or loss of energy nearly every day.
6. Feelings of worthlessness or excessive or inappropriate guilt nearly every day.
7. Diminished ability to think or concentrate, or indecisiveness, nearly every day.
8. Recurrent thoughts of death, recurrent suicidal ideation without a specific plan, or a suicide attempt or a specific plan for committing suicide.

Building on the approach to categorical diagnostics I have sketched out so far, which has been the standard since the DSM-III (APA, 1980) and ICD-10 (WHO, 1993) and still applies to most instances of disorder categorisation, a new feature has been introduced in the recent editions of the diagnostic manuals: making dimensional ratings part of categorical diagnostics.[13] The general idea behind dimensional ratings

13 These changes were introduced following the increased interest in psychiatric research in thinking of at least some psychopathological features as occurring on a spectrum. Proposals in this vein were made early on for personality disorders (Trull and Durrett, 2005) and psychosis (Esterberg and Compton, 2009; Cuthbert and Morris, 2021), for example, and were

is to evaluate the presence of at least some psychopathological features on an ordinal severity rating scale rather than by simple presence or absence. Dimensional ratings have been introduced as mandatory in the evaluation of diagnostic criteria for some mental disorder categorisations in the DSM-5 (autism spectrum disorder, intellectual disability) and as optional for others (primary psychotic disorder and personality disorders), and they are mandatory in some disorders categorised by ICD-11(WHO, 2019) (autism spectrum disorder, personality disorder) and optional for others (primary psychotic disorders).

While the basic idea is always the same, the use of dimensional diagnostics can take different forms. In some instances, dimensional rating systems are simply used as add-ons to the specification of present symptoms, for example whether the delusions present are clinically mild, moderate, or severe. In other cases, as in the personality disorder diagnostics in ICD-11, significant changes accompanied the implementation of dimensional diagnostics. In the case of personality disorder diagnostics in ICD-11, for example, the change was a deflation of the rich personality disorder taxonomy present in ICD-10 in favour of one general personality disorder category to be specified in its severity (no difficulty, mild, moderate, or severe) based on dimensional ratings of the patient's personality and social functioning, which is then further specified by selecting from a list of pathological personality features present in the case at hand.

To get a better idea of what the inclusion of dimensional diagnostics in disorder categorisation may look like, let me consider the personality disorder module from the DSM-5 (APA, 2013, p. 761) in more detail. To be diagnosed with a personality disorder, the patient has to show "moderate to great impairments in personality functioning" in relation to him//herself and others and at least one pathological personality trait in addition to a relative stability of the condition across time (>2 years) and across life contexts (intimate relationships, work, school, etc.).

The impairment in personality functioning is assessed by rating the patient on four dimensions (identity, self-direction, empathy, and intimacy) whose scales have five levels of severity (no impairment (0), some impairment (1), moderate impairment (2), severe impairment (3), and extreme impairment (4)). For each level of impairment on every dimension, descriptions of three typical features of patients who would be rated in this way are supplied. Someone may be assessed to be severely impaired (3) on the empathy scale, for example, if they are "hyper attuned to the experience of others, but only with respect to perceived relevance to self" (APA, 2013, p.

---

adopted in one way or another by relevant research movements in the field, such as the National Institute of Mental Health RDoC Project (NIMH, 2013) or the HiToP Research Consortium (Kotov et al., 2017). To discuss the scientific and clinical motivations to push for a dimensional understanding of mental disorder is beyond the scope of my project. For discussion of these, see, e.g., Helzer et al., 2009; Krueger and Bezdjian, 2009; Adam, 2013; Reed et al., 2019.

776), in accordance with one of three descriptions of the level of impairment in this domain. In addition to this dimensional rating, at least one of five proposed pathological personality traits (negative affectivity, detachment, antagonism, disinhibition and psychoticism) have to be determined to be present in the patient, and they may be further specified by choosing specific facets of these traits that are listed in the diagnostic module. To expand on the example of someone impaired in empathy, one may often also identify as present the trait of antagonism, defined as "behavior that puts the individual at odds with other people, including an exaggerated sense of self-importance and a concomitant expectation of special treatment, as well as a callous antipathy towards others, encompassing both an unawareness of others 'needs and feelings and a readiness to use either in the service of self-enhancement'" (ibid., p. 780). In the end it has to be decided whether the attributed combination of impairments and personality traits in a patient matches with a personality disorder category (now also specified in terms of personality functioning impairments and traits). If so, this category may be attributed to the patient. If not, the patient may nonetheless be diagnosed with a personality disorder that does not fall into one of the typical categories.

To explore one of the examples of the dimensionally adapted format, let me present the proposed diagnostic criteria for a schizotypal personality disorder (APA, 2013, p. 769):

A. Moderate or great impairment in personality functioning, manifested by characteristic difficulties in two or more of the following four areas:
   (a) **Identity:** Confused boundaries between self and others; distorted self-concept; emotional expression often not congruent with context of internal experience.
   (b) **Self-direction:** Unrealistic or incoherent goals; no clear set of internal standards.
   (c) **Empathy:** Pronounced difficulty understanding impact of own behaviors on others; frequent misinterpretation of others' motivations and behaviors.
   (d) **Intimacy:** Marked impairments in developing close relationships, associated with mistrust and anxiety.
B. Four or more of the Following six pathological personality traits:
   (a) **Cognitive and perceptual dysregulation** (an aspect of **Psychoticism**): Odd or unusual thought processes; vague; circumstantial; metaphorical; overelaborated; or stereotyped thought or speech; odd sensations in various sensory modalities.
   (b) **Unusual beliefs and experiences** (an aspect of **Psychoticism**): Thought content and views of reality that are viewed by others as bizarre or idiosyncratic; unusual experiences of reality.

(c) **Eccentricity** (an aspect of **Psychoticism**): Odd, unusual, or bizarre behavior or appearance; saying unusual and unappropriated things.
(d) **Restricted Affectivity** (an aspect of **Detachment):** Little reaction to emotionally arousing situations; constricted emotional experience and expression; indifference or coldness.
(e) **Withdrawal** (an aspect of **Detachment**): Preference for being alone to being with others; reticence in social situations; avoidance of social contacts and activity; lack of initiation of social contact.
(f) **Suspiciousness** (an aspect of **Detachment**): Expectations of – and heightened sensitivity to – signs of interpersonal ill-intent or harm; doubts about loyalty and fidelity of others; feelings of persecution.

Let me now move from the presentation of the two diagnostic formats in themselves to their relationship to each other. As with the components of diagnostic information-gathering, the diagnostic case formulation and the categorical diagnosis can be brought into a functional relationship to each other. Although the APA (2013) makes no explicit statement on the relationship between the two, it provides some remarks regarding what is necessary and sufficient to make a psychiatric diagnosis and introduces the notion of diagnostic "clinical judgement" in this context. Together these elements allow to reconstruct the relationship in question.

The American Psychiatric Association (APA) states of the categorical diagnosis that "it is *not sufficient* to simply check off the symptoms in the diagnostic criteria to make a mental disorder diagnosis" (APA, 2013, p. 19; my emphasis) but that "the relative severity and valence of individual criteria and their contribution to a diagnosis require clinical judgment" (ibid). Clinical diagnostic judgement, however, is a capacity whose acquisition "requires clinical training" enabling a psychiatrist "to recognize when the combination of predisposing, precipitating, perpetuating, and protective factors has resulted in a psychopathological condition [...]" (ibid.).

In this description of diagnostic clinical judgement, it is necessary to assess which aspects of a patient's experiences and behaviours qualify as symptoms and signs and what level of severity they manifest. Both requirements come down to what I described earlier as the clinical reasoning process through which psychiatrists develop their case formulation. The case formulation sets down the results of the psychiatrist's analytic work on the diagnostic information, which suggests that specific complaints do or do not have psychiatric symptom/sign value. In turn, this attribution of symptoms and signs whose justification is given in the case formulation allows for a quicker application of the proposed diagnostic categories and helps justify their application. The profiles of categories, consisting of lists of signs and symptoms and their severity, can be compared to those psychopathological conditions that the diagnostic case formulation suggests are present in the patient, and a diagnostic category can be chosen.[14] If presented alongside the categorical diagnosis, the case formulation thus makes transparent the reasons for which a specific categorical choice was made and so stands in a justificatory relationship to the categorical diagnosis. This relationship is illustrated in Figure 6.

Figure 6: Relationship between diagnostic case formulation and categorical diagnosis. Arrow indicates a background information relationship

More than merely being a plausible and elegant option to make sense of the coexistence of the case formulation and the categorical diagnosis, this way of understanding their relationship helps to avoid puzzles that arise otherwise. Consider that this relationship does not hold. Why, then, should the psychiatrist invest effort in a diagnostic case formulation that allows her to individuate complaints as being (non)-psychopathological symptoms and signs, if the outcome did not inform the diagnostic category choice? The diagnostic case formulation would seem pointless. If this were true, however, the question arises of how else the psychiatrist would learn about the presence of signs and symptoms. If there is diagnostic judgement at work that, as the APA requires, consists of more than just "checking" symptoms, then what is this process that basically does the same work as the diagnostic case formulation but that an opponent of my proposal would have to claim to be something different? And if there were something that did this work for a second time, why has it not replaced the diagnostic case formulation as part of a comprehensive diagnostic proposal? It appears *prima facie* that rejecting the proposed relationship between the diagnostic case formulation and categorical diagnosis would only generate new

14 To illustrate this, one might recall my previous example of the patient who complained about his sleep problems, which I used to indicate what proper evaluation may look like and why the information it produces can be crucial. In both cases, the reported complaint is superficially the same and may one think of the presence of the symptom of insomnia. We then saw that for good reasons the complaint will be evaluated to be a non-psychopathologically relevant complaint in the one case, but to qualify as insomnia in the other. In both cases, however, the sleeping problems and the explanation found for them by the clinician would appear in a case formulation for the patient, but in one instance addressed as psychiatric symptoms, in the other instance addressed only as disrupted sleep due to bad sleep hygiene.

puzzles, rather than solving or helping to avoid any. Therefore, I will assume that the relationship as presented here is adequate.

In accordance with my presentation of and remarks on the output of the diagnostic proposal, the overview flowchart must be modified as follows:

Figure 7: Modified flowchart of stepwise psychiatric assessment as developed in this section. Vertical arrows connect steps in the process; horizontal arrows indicate the influence of background information.

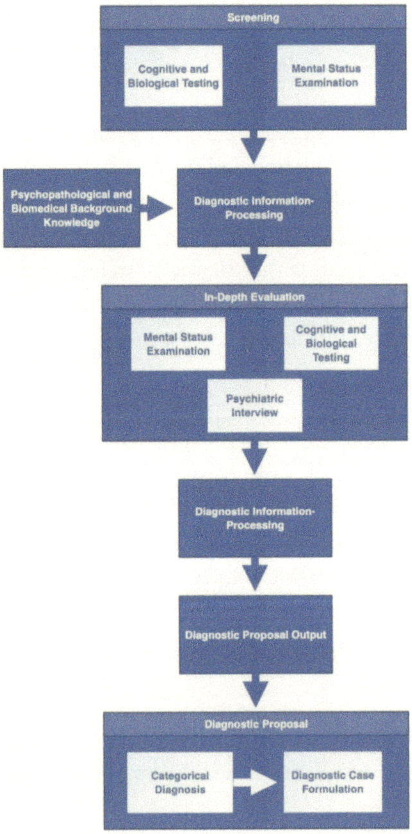

1.3 Diagnostic Information-Processing

Now that we have unpacked diagnostic information-gathering as well as the diagnostic output, the remaining aspect of the diagnostic process to be considered is diagnostic information-processing. I have saved the discussion of this aspect of the diagnostic process until last because it provides an interesting problem that makes for a good transition from the descriptive task of this chapter (to present a picture of psychiatric diagnostics) to the explanatory task of the next chapter: to spell out the diagnostic reasoning process that governs psychiatric diagnostics and to answer the Methodological Question.

So far, I have discussed what happens in diagnostic information-processing in only a very abstract manner. In section 1.1, I indicated how the screening guides the psychiatrist towards the decision of which potential psychiatric symptoms the patient needs to be assessed for, and how the psychiatrist uses background knowledge plus the variety of diagnostic tools at hand to carry out an in-depth evaluation of the patient, leading to an inference as to the presence (or absence) of psychiatric symptoms. In 1.2, I mentioned that in the inferential step from the in-depth evaluation to diagnostic conclusions, the resulting knowledge about the patient's psychopathological condition is used to set up the diagnostic case formulation, which is intended to explain the patient's condition and to generate an adequate diagnosis of mental disorder.

This abstractness in describing diagnostic information-processing results from the fact that the processing steps in question are usually discussed either in precisely this type of abstract and rather uninformative way, or in terms of single case examples that do not provide a generalisable framework useful for understanding what happens at this step of the process in general. Why this may be the case is puzzling. From conversations with clinicians and from my review the literature, it appears to me that clinicians learning to diagnostically assess patients learn and correct their diagnostic reasoning on a case-by-case basis. That is, they learn by looking at and working with single cases or small clusters of cases (i.e., patients with this or that pathology) rather than making use of a general framework governing what it means to process diagnostic information. Although such a general approach is perhaps tacitly picked up and skilfully exercised by clinicals who have been educated mostly via single cases and small clusters, the canonical presentations of psychiatric diagnostics contain no explicit reference to *how* diagnostic information-processing is supposed to take place in a generalised format.

If one looks for work on what happens in diagnostic information-processing, proposals can be found, but these proposals are not descriptively stating what can be generally agreed to happen in diagnostic information-processing. Rather, these are already theoretical proposals for how to understand diagnostic reasoning based on the rough commonsensical descriptions we have of it and how, given the inputs and outputs to this stage (and maybe some experimental data), we should understand diagnostic information-processing. These proposals try to provide a theoretical framework to explain what kind of belief-forming procedure takes place between the various stages of the diagnostic process. In so doing, however, they end up making a proposal as to what method is at work here. Examples of such proposals were briefly mentioned in the Introduction of this thesis, such as the phenomenological proposal (Fuchs, 2010; Parnas, Sass, and Zahavi, 2013) involving the idea that the reasoning process leading to the attribution of a disorder diagnosis is a form of Gestalt recognition. This is apparently not a commonsensical description of what psychiatrists do, but rather a specific form of theorizing about what they do. It is a part of an answer to the Methodological Question rather than a descriptive presentation.

Spelling out the diagnostic information-processing in a less vague but still generalisable manner seems not to be a task that can be addressed descriptively, thanks to the lack of consensus-building discussion on the topic within descriptions of psychiatric discourse. It seems that by looking at all we know about the diagnostic process as it is described here, proposing an understanding of what process is taking place in diagnostic information-processing is an explanatory rather than a descriptive task. Therefore, the descriptive work in this chapter is now complete. To address the question of how diagnostic information-processing should be assumed to take place becomes an interesting problem that we can look forward to seeing answered as part of the Methodological Question. Bearing in mind the question of how diagnostic information-processing should be thought to take place, considering all our descriptive knowledge of diagnostic core procedures, I will proceed towards addressing this and other questions. For now, I will briefly recap the main conclusions reached in this chapter.

1.4 Conclusion

In this chapter I have presented an overview of the core practices of clinical psychiatric diagnostics, to provide a descriptive baseline understanding towards which I can orient my approach to providing an answer to the Methodological Question. I started with the picture of diagnostics being a three-stage process involving diagnostic information-gathering, diagnostic information-processing, and, finally, the output of a diagnostic proposal. I unfolded each of these steps in the course of the chapter.

First, I discussed the diagnostic assessment and divided it into two further steps: the screening and the in-depth evaluation. I discussed the methods that typically provide the core of the psychiatric diagnostic proposal: the diagnostic interview, the MSE, and potential cognitive or biological testing.

Second, we proceeded to consider the results of the diagnostic assessment: the diagnostic case formulation and the disorder diagnosis proposal. I provided examples for both formats of diagnostic output and clarified how their relationship should plausibly be understood. Specifically, I claimed that the diagnostic case formulation presents the reasoning process leading to the psychiatrist's conclusions regarding the presence of certain symptoms, thus serving as the informational base for providing a symptom criteria-based disorder diagnosis.

Finally, I discussed the obstacles to addressing the aspect of diagnostic information-processing in psychiatric diagnostics, which is usually either described only in rather vague terms that can barely be considered to truly unpack what is going on, or else considered only in terms of single instances of diagnostics that do not provide a generalisable understanding of diagnostic information-processing comparable to the detail in which the other steps in the diagnostic process were spelled out. I suggested that as a result, the task of coming up with an understanding of diagnostic information-processing forms part of the process of generating an answer to the Methodological Question, rather than falling under the descriptive aims of this chapter.

Now that I have provided a description of the core procedure of psychiatric clinical diagnostics and thus established a baseline for what my methodological proposal must explain, we can proceed to the next step. This will be, in Chapter 2, to present a methodology for diagnostic modelling, which in Chapter 3 will then be argued to apply to the picture of psychiatric diagnostics being painted here.

2. Modelling, Qualitative Models, and Model-Based Diagnostics

This chapter will focus on the topic of modelling. It plays the important role of proposing an understanding of modelling that, as I will argue in the next chapter, maps onto the previously established picture of psychiatric diagnostics. This in turn will establish my proposal, the *model-based account of psychiatric diagnostic reasoning*, as an answer to the Methodological Question.[1]

While a whole chapter on modelling may seem excessive at first, it is crucial. It is crucial to give space to development the framework for modelling that I intend to apply to psychiatric diagnostics, because the proposed understanding of modelling has to meet specific requirements, mentioned in the Introduction to this thesis, to provide a methodology of modelling that, if successfully applied to psychiatric diagnostics, provides a satisfying answer to the Methodological Question. To recap, it needs to provide a description of the method assumed to be at work in psychiatric diagnostics, it has to provide an understanding of the rationale for the method to operate the way it does, and it has to speak to us about why and how the conclusions of the method may be deemed justified. Only when an understanding of modelling that can address all these points is established will an attempt to map the proposed method of modelling onto psychiatric diagnostics be able to yield a qualified answer to the Methodological Question. To generate this fully developed account of modelling, an entire chapter is required. Let me next sketch how the chapter is set up.

I begin this chapter by presenting a description of the type of modelling that I take to be realised by psychiatric diagnostic reasoning – namely, qualitative diagnostic modelling. To this end, I first (2.1) provide a general a general account of modelling, distinguishing it from other kinds of theorising based on contemporary debates in philosophy of science. Next (2.2), I introduce a specific format of modelling, qualitative modelling, as well as (2.3) a certain application of modelling, diag-

1 I have already begun to think about a model-based account for psychiatric diagnostics in Kind (2023). As the reader familiar with my previous work will note, the understanding of the type of modelling I discuss changed and evolved since my earlier reflections on the topic though I still take d Godfrey-Smith's (2006) and Weisberg's (2007; 2012) work as a starting point.

nostic modelling. After providing this description of the relevant type of modelling, the remainder of the chapter focuses on three things. First (2.4), I analyse the inferential strategy used in diagnostic modelling to provide an epistemic understanding of the rationale behind diagnostic modelling (why this kind of modelling proceeds as it does), and spell it out in terms of what I call the constitutive-indicator strategy. Second (2.5), I discuss the types of inferences executed by following the constitutive-indicator strategy, which I argue to be *inferences to the best explanation*, *apophatic inferences*, and *inferences to unintelligibility*. Third and finally (2.6), I discuss to what extent these inferences occurring in diagnostic modelling may justify its conclusions. I conclude (2.7) with a brief summary of the chapter.

2.1 Modelling

My general understanding of modeling is a slightly modified version of Godfrey-Smith's (2006) and Weisberg's (2007; 2012) accounts, with the latter building upon the former. Their accounts were derived from case studies in evolution and population biology and informed by previous debates on modelling, mainly in the philosophy of physics and economics (e.g., Cartwright, 1983; Wimsatt, 1987; Giere, 1988; Morgan & Morrison, 1999). Currently, their view is not only highly plausible, it is also the most comprehensive and detailed account of modelling as an epistemic practice in the philosophy of science. Therefore, their account is a strong candidate for determining whether a certain epistemic practice, such as psychiatric diagnostics, should be classified as modelling.[2]

Godfrey-Smith's and Weisberg's main idea is that theorists developing and using models (i.e., modellers) follow a particular strategy of theorising to develop theoretical models of empirical systems. They call this strategy the *indirect strategy of representation* (from now on ISR). A theorist following this strategy engages in a three-step procedure. First, they set up a theoretical structure based on limited initial information about the target system and assign aspects of this structure to an element of the targeted real-world system. Second, the theorist investigates the properties of these theoretical structures to learn about its dynamics in order to predict its future states and outputs. Third, the theorist compares the findings of the structure's properties to the behaviour of the real-world system(s) that the theoretical structure was intended to target and judges whether the structure can be used to satisfy the theorist's epistemic interest in the system, for example by predicting changes or simulating entire processes taking place within it.

2 Another feature of this account making it attractive for dialectic reasons is that it is free of controversial commitments regarding the ontology of models and theories of model representation (Frigg and Hartmann, 2020).

If it turns out that the structure enables the theorist (and other competent users) to make this inference with sufficient precision to meet their needs, the structure is accepted for usage and considered to be a credible model.[3] If the structure does not meet the pragmatic criterion of being useful to the modellers' aims, it will be rejected and either another structure is set up or a modified version of the already tested structure receives a second round of analysis and comparison to achieve credibility. The individuals following these steps are automatically considered to be modellers.

To contrast modelling (i.e., following the ISR) with other forms of theorising, Godfrey-Smith and Weisberg introduce an approach to theorising that they call the *abstract direct strategy of representation*. While following the ISR procedure is modelling, following the abstract direct strategy of representation is supposed to result in data-driven theorising, which in their understanding is distinct from modelling. A theorist following the abstract direct strategy of representation proceeds as follows. They begin their pursuit of a theoretical structure that targets real-world systems by generating and collecting large amounts of available data about the system(s) of interest. This way they address it *directly* before they begin to theorise. Based on the large amounts of data they collected, they try to determine which properties of a system appear to be essential to account for other properties of the system and set up a theoretical structure based on this judgement. In this process they abstract the rest of the data not needed for this purpose. In the end, they arrive

3 Credibility is a matter of pragmatics rather than truth (in the sense of faithful/complete representation), and it is the central goodness criterion applied to evaluate models (e.g., Sargent, 2010; Truran, 2013). For a model to be credible it is neither *necessary* nor *sufficient* for it to be a faithful and complete representation. It is not even typical. It is not necessary because many if not all predictive models and simulation models are highly idealised, but they are nonetheless sufficiently predictive and/or simulatively accurate to be used in alignment with the modeller's interests and therefore credible. Faithful and complete representation is not sufficient because credibility is a pragmatic matter, and a faithful representation may under some circumstances not be the right tool to archive the aims of a modeller. If, for example, a modeller wants *one* model that makes a prediction (with some margin of error) about multiple similar but in many regards different systems, it seems possible that no faithful representation of any of these systems – and there can be *one* faithful representation of several different systems – could provide the modeller with a "model" credible for the task of making the desired predictions across this systems, while it may be that a model that contains idealisations is in fact well suited to the task. Note that the latter illustrates a *contingent* and not a *necessary* tradeoff between precision and credibility resulting from the interests of the model. I take this to be uncontroversial, whereas claiming a necessary tradeoff relationship between precision and utility would be a highly controversial claim (Odenbaugh, 2003; Orzack, 2005, 2012; Matthewson and Weisberg, 2009). Finally, representational fidelity or completeness is *atypical* for what we call models, since in speaking about models we do not consider models to be anywhere close to being faithful complete representations of their target, but at best *partially* true representations of their targets (e.g., Da Costa and French, 2003).

at a theoretical structure meant to represent the target system's properties and their relations faithfully.

The crucial difference between both ISR and the abstract direct strategy of representation is that while the representational features of both theoretical structures are indirect in the sense that they end up proposing a theoretical structure serving as a vehicle for reasoning about a targeted system, ISR is indirect in an additional sense. It takes a deliberate extra step of setting up a theoretical structure to stand in for the system of interest and investigating this structure to learn from it, and *only then* relates this setup structure to the target system in order to evaluate one against the other. The abstract direct strategy, on the other hand, begins by directly collecting vast amounts of data about the intended target system, investigates the data, abstracts the materials that are unhelpful for arriving at a representation of the features of the target system that the theorist is interested in, and then sets up the theoretical structure on the basis of the retained data.

Now that the difference ISR and the abstract direct strategy is clear, let me add that the difference appears gradual rather than categorical. It is rare that a theorist sets up a theoretical structure meant to target a real-world system without *any* knowledge of the system. At a minimum, the theorist must have knowledge of the existence of the system, knowledge that gives reason to be interested in it, and some assumptions about it that leads them to set up one or another structure to target it. On the other hand, even those theorists who engage in some form of data-driven theorising about systems have a disciplinary background that provides specific structures typically used for theorising. Likely, those structures will be applied to analyse data that do not in themselves tell us how to order and analyse the data or make inferences about the real-world system based on this data.

Equipped with an overview of the principal strategy followed by modellers, let us go into more detail regarding each step of the modelling process: how we construct models, how we analyse them, and what we can learn from them about reality.

2.1.1 The Construal

The first step in this process is constructing the model system that is meant to target a real-world system. This step might be guided or inspired by existing theoretical sources providing full or partial structures for the model, by the limited knowledge about the target system, and by presumptions about principles that may govern aspects of the system or require the modeller to draw on previous experiences from (un)successful attempts to model similar processes or systems. Bringing all these sources of inspiration in play has been called the "*art*" or "*know-how*" aspect of modelling for which no manual exists (Morrison, 1999; Godfrey-Smith, 2006). It requires experience and expertise in modelling. The step of construal itself is characterised by four aspects. These aspects are not considered to be steps that have to be carried out in chronological order; they are interdependent and must be considered simultaneously by the modeller. The four aspects are: finding a structure for the model, assigning the model to a target, determining the scope of the model, and setting up its fidelity criteria.

Structure: To construct a model, a theorist must select a theoretical structure via which they present the model. Such a theoretical structure may be quantitative or qualitative. It could consist of graphical representations, mathematical formulae, or interrelated propositions expressed in written text. Because one and the same intended underlying model could be expressed in different theoretical structures (i.e., a box-and-arrow diagram versus a formula versus a verbal description), while we assume each of these expressions to represent the same model, the chosen structure put forward will not be *the model*; it will be only one possible *description* of the model.[4]

In choosing a description, modellers will often be attempting to choose the one allowing them to capture the intended model's elements and relationships as precisely as possible. Moreover, the modeller will consider how reasonably a chosen structure will be able to be exploited in targeting the specific aspects of real-world system the model is intended to target (see "Assignment" and "Scope" below); what inferences to make about the system the structure is meant to enable (see "Fidelity" below); and what resources the modeller has available in order, later on, to compare a given model to the real-world system (see "Model/World Comparison" below). To come up with a model structure meeting all these requirements, the modeller can call on various sources.

One source employed might be the modeller's intuition, fuelled by methodological training and ideas about the target system. Another common source of inspiration for model structures are existing model structures, from the same or other branches of science, that have been used to model similar phenomena.[5]

4 Recognising the difference between the use of model descriptions rather than the actual model is not only plausible; it is also helpful for avoiding metaphysical questions. It is plausible, because not making this distinction would have implausible consequences. We would have to say, for example, that graphical illustrations of models that are also mathematically presented in scientific papers are not two descriptions of the same model but two different models. That is surely not what is intended by the scientists. It sounds more plausible that both the maths and the graphics describe the same model. Moreover, by admitting that all we are really dealing with in the process of modelling are different forms of model descriptions, we can avoid deep metaphysical discussions about models, such as whether in the end all theoretical models are mathematical and thus whether it is correct to speak of models that are not mathematical. Such problems can be avoided by the plausible assumption that what we encounter, modify, and handle in modelling are just descriptions, whatever the deep metaphysical truth about models might be.

5 The first kind of model reuse is called cross-contextual modelling (Knuuttila and Loettgers, 2016). A famous example are the Lotka-Volterra equations, first proposed to model autocat-

Finally, model structure may be based not on existing models but on assumptions articulated in theories that seem promising for addressing the relevant aspects of the target system. However, while reusing a model structure is often fairly straightforward, using a theory to come up with a model structure can be tricky. Sometimes theories make claims that are too abstract and do not in themselves provide specific structures for their application. Rather, an applicable structure has to be engineered based on the theory's principles.[6] Sometimes neither *one* theory nor *one* model provides the modeller with a coherent structure that appears a plausible candidate to map onto the regularities of interest in the system. In such cases, the modeller must draw on multiple models and theories providing partial structures or a basis for such structures, and this collection is then 'pieced together.'.[7]

Assignment: The assignment process accompanying the choice of the model structure specifies the model's target systems(s). In other words, it determines what the model is a model of, as well as which parts of the model structure target which aspects of the target system. Depending on how the model structure is specified, this assignment process might come more or less naturally. The assignment is more noticeable when the specification of the model's structure itself contains obvious hints – for example, if this model description contains symbols referring to aspects of the real-world system they are intended to be assigned to. For example, think of the small pictures of animals in a typical presentation of the "tree of life" model of evolutionary history, or the boxes of a model from a neuroscience textbook with brain areas names in them. Things are less straightforward, however, in models that are expressed in purely quantitative terms. These models need a more explicit articulation of their intended assignment. This problem is often solved by modellers through

alytic chemical reactions (Lotka, 1910) and later applied to model predator–prey interactions (Lotka, 1925; Volterra, 1926) and economic fluctuations (Goodwin, 1967). The second kind of model reuse has been described in terms of hub models (Godfrey-Smith, 2009). For example, suppose you have a detailed understanding of how one particular trait became selected in an evolutionary process. In that case, you might attempt to apply the same structure to model another trait's selection.

6 For an example, think of social network theory. This theory does not provide you with a specific network model to apply, but instead tells you how to set up a structure that complies with the theory. Or, to take a classic example, classical mechanics does not provide you with a model of a pendulum, but gives you the tools to develop a model structure for a real-world pendulum by providing a framework for taking different real-world factors (e.g., friction) into account in attempts to develop a model (Giere, 1999).

7 A nice discussion of several of these "puzzling examples" can be found in Boumans (1999) in the context of modelling in economics, where modellers attempt to integrate different sources (e.g., economic models, phenomenological laws, and assumed economic "laws") into business-cycle models.

conventions as to how elements of model descriptions are meant to map onto aspects of the model system. In physics, for example, there are established lists of constants. Everyone familiar with classical mechanics looking at a mechanical process model knows that *F* is assigned to force. In medical and social sciences, it is common knowledge that *n* stands for the number of subjects in a sample.

Scope: The third component of the model's construal is selecting its scope to determine which aspects of the phenomena will be targeted. This interest-guided process is the other side of the assignment process. The assignment establishes the intended mapping of the model onto the real-world system by telling us what real-world system the model targets and what parts of the model are meant to relate to selected aspects of the target system. The model's chosen scope determines the target systems' aspects to be targeted by the model and which aspects of the target system are left aside. The scope is usually determined by the modeller's interest and the presumptions about the aspects of the real-world system that will be relevant to achieve a credible model. For example, suppose you know that in ecological modelling of population growth, considering the amount of prey is an essential predictor in many models. In that case, as a modeller you might decide to take it into account for your own model of population growth.

Fidelity: The final aspect of the construal is the stipulation of fidelity criteria. They define adequacy conditions for models. Fidelity can be divided into two types. First, a model's *dynamic fidelity*, determining how similar the output of a model (its prediction) has to be to the real-world system's output to be considered credible. This criterion may take a numerical value and ordinal positions or take the form of quality space, and in the first case it is often provided with an error tolerance value regarding its outputs. The second kind of fidelity is *representational fidelity*, determining to what extent the model structure allows for simulations of changes that also occur in the aspect of the real world system to which its elements are assigned. This is considered important if the purpose of the model is not only to predict but also to track causal pathways in the modelled system, usually leading to higher credibility requirements when it comes to representational fidelity.

2.1.2 Analysis of the Model

The second stage of the modelling process is the investigation of the model system itself. In this step, modellers familiarise themselves with the model structure they have developed and with its dynamics and predictions. In other words, they learn what elements are present in the model and what changes to which elements of the system lead to what changes in other elements.

This analytical step is logically autonomous from the model's application to a target system. For now, the modeller is concerned only with discovering regularities in the model structure. As long as the model passes the later step of model/world comparison, these regularities may suggest interesting options for using the model for epistemic and practical purposes in application to the targeted real-world system. Things that might be discovered in such analysis that may be interesting later include, for example, surprising relationships between elements in the model discovered in a simulation that might be useful in predicting changes in the targeted real-world system, or in guiding strategies to develop interventions to control changes taking place in this system. Moreover, successful models may even give rise to principles that might not only be true in the one specifically modelled system but might also turn out to be generalisable to other systems. As Weisberg puts it: "Even where the model is inspired by a real-world system, what the theorist finds out about it is distinct and usually more general from the system which inspired it" (2007, p. 19). Examples of such principles developed from a single narrow, targeted model include the "Volterra Principle", in which general pesticides (i.e., an intervention that kills both predator and prey) increase the relative proportion of the prey (Roughgarden, 1979), and "Dunbar's Number" (Dunbar, 1992; Hernando et al., 2010), stating that human groups larger than 150 will not be sustainable based on personal relationships. A feature of model analysis is therefore not only the promise of discovering regularities that might hold in the specific system it will be evaluated against, but also the potential discovery of more general principles applying to other relevantly similar systems.

2.1.3 Model/World Comparison

In the last step of the modelling process, the results of the analysis are compared to the real-world system to see how well the model's predictions and simulations map onto the aspects of the system targeted by the model. In such a comparison, the modeller compares the model not directly to the targeted system but instead to a description of this system. This description needs to represent the system's relevant features in a format that matches the format of the model structures, so that a comparison is possible.

The need to craft a theoretically adequate description of a system in order to bring it in touch with models has been helpfully identified in philosophy of science by Nancy Cartwright (e.g., 1989, p. 133). She introduced the difference between a *prepared* and an *unprepared* description. The unprepared description "contains any information we think relevant in whatever form we have it available [...]. We write down whatever information we have" (ibid.). However, such a description does not usually allow a model to be assessed against it. For this we need a prepared description instead. In such a description, "we prepare the phenomenon in a way that will bring it into the theory" (ibid.) – or in our case, the model. If, for example, we have created a mathematical model of the changes of temperature in the course of a chemical reaction, the description of the system (the reacting compound) will have to provide a representation of the target system's modelled aspects in a numerical form that makes its states and changes relatable to the proposed mathematical structure of the model. This means that if the temperature of the compound is modelled in terms of degree Celsius, the prepared description of the system's heat states will also have to be given Celsius, to allow for assessment of the model.

If such a prepared description of the systems is provided, the theoretical model is compared to it, and the modeller decides to accept or reject the model in light of the results of the comparison. The model will be accepted if the model matches the behaviour of the system well enough to meet the modeller's interest in predicting or simulating the system, and if it can therefore considered to be a credible model of the system. The "good enough" status is determined on the basis of the previously stipulated fidelity criteria. If the model mismatches the system – that is, the comparison reveals that the model does not meet the initially formulated fidelity criteria – the model will be rejected. However, if the modeller does not intend to end their efforts at this point, such a mismatch can itself be used to modify the model structure in an attempt to come up with a model that better fits the targeted system. If this route is taken, it requires another round of analysis and comparison with the model's target to establish whether the resulting model might be more acceptable. If not, more fitting work or the invention of a totally different model structure might be the path the modeller has to take.

So far, I have presented the basic outline of the modelling that I will apply to psychiatric diagnostics. In the next section I will flesh out a variation of one aspect of the modelling procedure – namely, the potential choice of a qualitative over a quantitative model structure, and how such a choice influences the modelling process and its results. Understanding these consequences will be important for our purposes here, because I will claim that models used in psychiatric diagnostics qualify as qualitative models.

2.2 Qualitative Models and Qualitative Modelling

Modelling as presented above can take place in the form of either quantitative or qualitative modelling. In this section I will explore qualitative models and modelling. This step will be important for my project, since I will propose that psychiatric diagnostics, if it is understood as a modelling process, should be considered to involve qualitative modelling. To argue this point, a good understanding of the features of qualitative modelling is needed.

What makes modelling qualitative or quantitative is the nature of the employed structure. In this context, the *structure* of the model refers to the elements of which a model consists and the relationships between them. These elements and the relations between them can vary in nature. Most examples of models discussed so far, with some exceptions such as the "tree of life", were cases of scientific modelling in which the model structures are quantitative; their structure consisted of elements and relationships that are mathematically continuous variables. Qualitative modelling, on the other hand, is a form of theoretical modelling in which aspects of real-world systems are represented in a discrete and symbolical manner, no matter whether the real-world system is considered to be continuous or not. The chosen values introduced for the purpose of being assigned to states of the modelled system will in qualitative models usually be of limited number – e.g., "present", "absent", "neutral" – rather than being continuous variables with a potentially infinite number of values. Likewise, the relationships take the form of qualitative values such as "increases", "decreases", or "irrelevant" rather than indicating a quantitative measure for the influence of one variable on another. As a result of such discretisation of values in the model structure, each value can be understood symbolically, as making a reference to a discrete state or condition in the system. The presentation of the resulting qualitative model can take various forms. Such models can be presented in a drawing accompanied by guidelines on how to interpret it, as the "tree of life" model may be; they can be presented via the box-and-arrow diagrams we find in textbooks; or they can be represented in the form of conditionals in propositional logic. Qualitative models can also be presented as a set of interrelated propositions that are expressed by means of natural language, specifying the elements and relationships in the model. This last format for construal of a model has been called a *propositional model* (Thomson-Jones, 2012).

A typical and philosophically interesting feature of qualitative model if compared to quantitative models is their higher degree of idealisation. This higher degree of idealisation takes place in two forms that are terminologically differentiated in philosophy of science as *Aristotelian idealisation* (Batterman, 2002) and *Galilean idealisation* (McMullin, 1985).

Aristotelian idealisation is introduced to a representational structure in the context of determining the scope of the model. In this kind of idealisation, decisions are made about which features of the real-world system are intended to be represented by the structure and which will be abstracted from. The more is intentionally left out, the higher the degree of Aristotelian idealisation. If, for example, I develop a model of the population dynamics in an ecosystem but knowingly ignore certain populations that I know to be present (e.g., smaller animals like insects), I engage in Aristotelian idealisation. While such forms of idealisation take place in any kind of modelling, it is typical for qualitative models to show a higher degree of it due to the more limited number of elements and relations that usually appear in them compared to quantitative models. Of course, qualitative models *could* in principle have infinitely many elements and relations with infinitely many discrete qualitative states of parameter and relations so that the degree of Aristotelian idealisation could be decreased, but in practice this would undermine one of the main reasons why many choose this way of modelling – namely, the computational simplicity that grants cognitive tractability to its outcomes.

Galilean idealisation takes the form of deliberate distortions to the representational structure that targets aspects of the real-world system, to address them in a simplified way. For this, the elements and relationships in the structure undergo simplification that intentionally reduce the complexity of the targeted features. An example of such simplification would be developing a model showing how employment and education influence the likelihood of being in a long-term relationship, in which the variables and their influences represent known real-world complexities in a simplified manner. This may mean, for example, that education is differentiated only in terms of what degree one has, ignoring educational performance differences among people with the same degree, or that the variety of existing employment situations is reduced to the opposition between employed and unemployed. Regarding the relations between these elements, simplification might take place when the model assumes that being employed leads to a higher likelihood of being in a long-term relationship but intentionally ignores further factors that would complicate modelling this likelihood relationship, such as temporal factors (e.g., how long someone is employed) that the effect might depend on. While this type of idealisation also occurs in quantitative as well as qualitative models, qualitative models usually introduce a higher degree of Galilean idealisation than quantitative models. The usually limited number of elements and discrete values favours lumping together real-world phenomena that are in principle separable into fewer variables to ensure computational tractability, and relationships modelled as discrete qualitative states impair the capacity to address more complex relationships amongst variables. Given that both types of idealisation, Aristotelian as well as Galilean, are typically highly present in qualitative models, they are typically not the first choice of modellers interested in maximising the representational fidelity of their models. However, in a context where representational fidelity is not a central requirement for the model's use, qualitative modelling can have beneficial applications.

A central benefit of qualitative models is that they are cognitively more traceable and, *vice versa*, that they can provide a framework for a cognitively realistic understanding of expert reasoning.[8] Indeed, qualitative models not only *can* do this

8 However, it worth adding that qualitative models might also be chosen if suitable quantitative data about the target system that a higher-fidelity qualitative model would require are not available, or when the system itself is so complex that its computational intractability

but they do: one of their primary roles in research and practice is to understand and support the expert reasoning. Research on expert reasoning has shown that experts tend to think about the features of systems they interact with in qualitative terms. When thinking about quantities, motion, space, time, causation, or frequency, practical experts often do so without using abstract mathematical methodological frameworks, instead sticking to the qualitative categories of folk reasoning (Forbes, 2008).[9] This idea has been most prominently developed in the research on "qualitative physics", a branch of cognitive science investigating the use of qualitative models by engineers and other technical experts thinking about artifacts and their functions (see, e.g., Bobrow, 1984; Falkenhainer and Forbus, 1991; Weld and De Kleer, 2013). Similar research can be found in attempts to understand expert reasoning in economics and engineering as a form of qualitative modelling (Farley, 1987).

Considering what qualitative modelling has contributed to our understanding of expert reasoning in other fields, it also *prima facie* appears to be a promising candidate for understanding how psychiatrists approach their diagnostic task. As in the case of other experts who think about most of their work in qualitative terms, we can plausibly expect that the same is true for psychiatrists, since psychiatrists do not *calculate* their diagnosis, instead thinking about diagnostic matters mainly in qualitative terms. This becomes clear whenever one listens to diagnostic discussions at case conferences or in a clinical setting, from diagnostic discussion in training literature (e.g., Wright, Dave, and Dogra, 2017), and also from research on clinical reasoning that uses think-aloud protocols to show that clinicians think in qualitative terms about their cases (e.g., Audétat et al., 2012).[10] More about this will be said in the next chapter.

To bring my discussion of qualitative modelling to an end, let me ensure that we move on with a clear idea of how this kind of modelling works by providing an ex-

---

could render qualitative models better suited to predict or simulate aspects of the system that the modeller is interested in. Some recent examples of this later case can be found in areas of science dealing with immense complexity – for example, in attempts to model marine ecosystems, where highly idealised but tractable working models find applications (e.g., Reum et al., 2015).

9 Note here that this understanding of qualitative theorising differs from the understanding of Weisberg's (2004) discussion of qualitative theorising in chemistry. While qualitative models as discussed here share many features that he discusses too (e.g., a high degree of idealisation and a typically restricted number of variables), qualitative models as discussed here are not numerical, whereas Weisberg explicitly states that in his understanding, the difference between "qualitative and quantitative models is not about the use of numbers; both types of models can be numerical" (Weisberg, 2004, p. 1071).

10 For more on the validity of the use of this method, see Durning et al. (2013). Together, these forms of evidence seem to me sufficient to support the claim that if diagnostic reasoning in psychiatry is a form of modelling, it should be expected to be some sort of *qualitative* rather than *quantitative* modelling.

ample, adopted from Forbes (2008). Think about the process of heating and cooling water in a vessel on a machine that may increase and decrease the temperature of the water – that is, a freezer-heater. Assume that this machine has three positive and three negative levels (freezing, cooling, chilling, warming, heating up, boiling), and you are interested in the model of the conditions for water to take one of three qualitatively described states called "solid", "liquid", and "vaporising/boiling". These states are considered to appear on an ordinal scale such that any change between "solid" and "boiling" must pass the state "liquid". To develop a qualitative model for this real-world system, you may set up a structure with two variables, one assigned to the freezer-heater (saying what its current setting is) and one to the water in the vessel (saying in which state the water is). Then you assign a number of potential the qualitative values of the water and the heater-cooler setting to the elements of the structure. Which would make six potential values for the element mapping on the settings and three for the element mapping on the water. Then a relational structure can be set up, making claims about which qualitative value of the parameter assigned to the freezer/heater may lead to an influence on the parameter assigned to the water in the vessel. Finally, this model and its implications might be compared to the real world by playing around with the freezer-heater settings. Imagine that after some revisions that allow our model to match the real-world phenomenon, our model tells us that if the freezer/heater is on "freeze", the water goes down the ordinal scale stepwise until it is "solid". In contrast, the water goes up the ordinal scale until it is "boiling" if the machine is on "cook". In all other states, and the water will move towards (or stay on) the ordinal scale value "liquid". Suppose that these predictions are what can be performed by the model. In this case, we can make an accurate prediction following our initial interest in the relationship of the machine settings and the state of the water. And suppose that the water's transitions between the qualitative states assigned to them are also predictable in the model. In this case, we will also be willing to assign representational fidelity to it. The qualitative model meets the purpose of the modelling process without the need to set up a quantitative model of the system.[11]

11 What has been said so far should make clear the principal idea behind qualitative modelling, rather than its boundaries. Qualitative models *can* be more complex, and AI researchers and mathematicians have worked out frameworks to give technically more rigorous representations of qualitative modelling through qualitative algebra (Forbes, 2008). I do not intend to introduce and discuss any specific formal framework to talk about qualitative models; the educated intuitive understanding of qualitative modelling that should result from the previous presentation will suffice. What follows is therefore only gesturing in the direction of relevant work in AI and mathematics. For concrete examples of proposals for formalised qualitative models of complex systems, one may look at the examples of two-valued models, employed to diagnose dysfunctions in aircraft engines (Abbott et al., 1987) or photocopiers (Bell et al., 1994). A framework for employing three-valued formalisations based on a positive (+), nega-

Now that the outlines of qualitative modelling and its specificities have been provided, let us turn to the next specification of modelling that is relevant to its application to psychiatric reasoning: the use of models for diagnostic purposes, in the form of diagnostic modelling. In the next section, I discuss this specific use of modelling, and in the next chapter I will argue that this is the type of modelling that is also realised by psychiatric diagnostics.

2.3 Model-Based Diagnostics – Normative and Error Models

Modelling as understood in ISR varies not only in terms of the formats of models used (qualitative versus quantitative), but also in terms of the epistemic aims pursued through its application. Modellers may have specific interests in exploiting the regularities of the model system, evaluating the (dis)similarities between model structures and real-world systems, and assessing the outcomes of these evaluations. For instance, models that effectively capture variable changes over time can be utilised to predict specific occurrences in the modelled system, simulate its processes, or guide interventions to achieve certain changes in the system. As discussed with regard to the model analysis step, the ability of such models to match the modelled systems can be leveraged for a variety of purposes, highlighting the importance of careful model selection and analysis in ISR. Here I want to discuss another use of models: to identify and classify irregularities in the modelled system. The practice of setting up and using models for this purpose is called *diagnostic modelling*. If, as I argue, psychiatric *diagnostics* is to be understood as a modelling process, to consider it diagnostic modelling seems *prima facie* a plausible candidate. To further assess this plausibility, in the next chapter I will present this idea in more detail. In other words, in the remaining chapter, I will discuss what diagnostic modeling is and examine in the next chapter whether or not psychiatric diagnostics should be understood as a form of such diagnostic modelling.

The basic idea of diagnostic modelling was proposed by Reiter (1987). Diagnostic Modelling enables the decision as to whether an error of a certain type occurs in a real-world system via a comparison between this system's actual performance (in terms of outputs or internal processes given certain inputs at some point in time) with a presupposed model of the system, which I will call the *normative system*

---

tive, (-), or zero (0) value (on an ordinal scale) to model physical systems on different levels of complexity can be found in de Kleer and Brown (1984). Moreover, proposals have been made for the formalisation of monotonic relationships between model elements (e.g., if A goes up, B goes up) as well for compositional relationships (e.g., if A goes up, B goes up, *iff* C goes down) and how change over time can be considered in a time series of a qualitative model (Forbus, 1984).

model.[12] This normative system model is "a model which can be used to simulate the normal work of the system in the case of lack of any faults" (ibid., p. 440). Depending on how fleshed-out this normative model is, different aspects of the system at hand may be compared to it to find errors in the system. Given the normative system model and the actual system it targets, the modeller can then evaluate whether the real-world system shows the normal operations assumed by normative system model – that is, whether they can initiate a diagnostic process. In this diagnostic process, the modeller makes a "comparison of the observed system behavior and the one which can be predicted with the use of the knowledge about [the] system model" (ibid., p 440), which I call the normative system model. This way, deviations can be identified between the system's actual behaviour and how the system should behave under certain input conditions if it works without errors. Recognised deviations can then be diagnosed as errors.[13]

Based on this general idea of Reiter's, I want to propose some variations of model-based diagnostics that a modeller may engage in and that may lead them to make diagnostic statements of different levels of granularity. These kinds of diagnostic modelling will be relevant for understanding psychiatric diagnostics as modelling.

The first and simplest way in which model-based diagnostics takes place is what I will call *normative-model diagnostics*. In this kind of diagnostic modelling, the way to arrive at the conclusion that the system is in error is to look only at the inputs the system receives and the outputs it provides compared with what would be expected to be the output under the same condition in the normative model. In this method, all insights about the system error – including identification of errors and the entire

---

12 Reiter (1987) calls the model used for the comparison the "system model". I include the term *normative* to emphasise the model's function.

13 If you are familiar with recent developments in psychiatric research, you may wonder how the role of normative models as described here relates to recent uses of normative modelling in psychiatric research (e.g., Marquand et al., 2019). While both approaches make use of normative models, they do so for different purposes. The role of normative models in the account of diagnostic modelling proposed here is to enable the identification of errors based on deviation from an accepted normative model. Normative modelling in psychiatric research is "a class of emerging statistical techniques useful for *understanding* the heterogeneous biology underlying psychiatric disorders at the level of the individual participant" (ibid., p. 1415, my emphasis). Briefly and non-technically, this is meant to be done by establishing a "mapping between behavioral, demographic or clinical characteristics and a quantitative biological measure, providing estimates of centiles of variation across the population" (ibid., p. 1416). This mapping can then be used to try to identify biological variations found in individual deviations from the normative model. While normative models used in diagnostic modelling serve to identify deviations present in the system, the relevant use in psychiatric research is to support the discovery of biological variations co-occurring in individuals who are already assessed as deviating from the normal functioning determined in the normative model.

basis of a classification of them – are the system's outputs, which simply inform the modeller that the system itself must somehow deviate from its inner working and do not presuppose a detailed model about the system's inner constitution and normal functioning. A modeller may be satisfied by learning that the real-world system is *somehow* in error. They may be able to classify and identify system errors based solely on inputs and outputs, but they do not have to.

Diagnostic modelling has the potential to be more detailed;[14] the normative system model is not restricted to treating the system as a black box and only look at inputs and outputs. It may, in addition, address internal processes and components of the diagnosed system, yielding finer-grained diagnostic statements that do not identify a type of error based exclusively on unexpected input-output findings. Diagnostic modelling can serve to differentiate and identify errors based on what is occurring within the system when the erroneous output takes place. Such a closer look enables a more differentiated approach to identifying an error, based on a better understanding of how the system produces this error. This may even enable the modeller to differentiate between two types of error in a system that are indistinguishable in terms of an input/output relationship, but that differ regarding the system states presumably responsible for the erroneous output. In the latter case, this would allow the modeller to differentiate a *prima facie* singular error phenomenon into two errors of the system, which may lead to the use of two different diagnostic labels for them instead of one.[15] This higher level of detail in diagnostics is worth

---

14 At this point, one may ask where these normative models come from. They have to be established by previous system modelling efforts, under conditions that for the modeller community interested in the system were assumed to be normal working conditions.

15 Why do I say it may lead to more than one label? It might be the case that a modeller may set up a taxonomy with a one-to-one mapping between models and labels, but this is not necessarily the case. There might be reasons to establish a many-to-one mapping instead – from various errors to one and the same label. Reasons to prefer a many-to-one taxonomy may, for example, be pragmatic. Assume that the overall purpose of the labels is to guide interventions to repair the system, and among the considered interventions one type of intervention works just as well to return a system to normal functioning for multiple errors that lead to a similar erroneous output. In this case it would be a practical option – that is, one that allows us to succeed in our task considering the commitments we make in the attempt to fulfil it – to introduce only one diagnostic error label for the purpose of interventions. As Zacher (2002) pointed out with the example of psychiatric diagnostic labels, there are many practical commitments at work in coming up with a diagnostic classification system: "Deciding what counts as practical is complicated. With respect to categorizing psychiatric disorders, we should consider many things, including but not limited to available treatments; potential management strategies; the effects of labeling; maximization of true positives and true negatives in identification; establishing within-category homogeneity for creating groups in experimental research; uncovering etiologic scenarios (especially for spectrum disorders); mapping time courses; predicting prognosis; achieving coherence with basic science in genetics, physiology, and psychology; being both clinically informative and easy to use; and

pursuing not only for the sake of more precision in classification in itself, and the epistemic interest that might be related to this, but also because it promises pragmatic benefits relevant for modellers with interventionist interests: Finer-grained, more informative classification may aid in choosing interventions to target the relevant deviations of the diagnosed system and so restore normal working, or suggest how to modify the system to compensate if some parts of it are broken beyond repair. To achieve more detailed diagnostic modelling of this type, another kind of model is needed in addition to the normative system model; I call it the *error model*.

Error models are derived from theoretical background assumptions about *how* real-world systems deviate from the normative model when showing certain erroneous outputs. Just as in the case of the normative model of a system that has to be pre-established before diagnostic modelling can take place, error models are developed from the variety of sources described in 2.1 by the diagnostic modeller or their modelling community, providing them with a repertoire of models addressing not only the inputs and outputs of the system but also additional features that enable them to identify errors. The use of such error models presupposes, of course, that the normative model in contrast to which these errors are meant to be identified makes assumptions about how the system components or processes addressed by the error model should behave normally. If the normative model meets this requirement and error models for a certain type of error exist, three different kinds of diagnostics that uses error models may take place: *error-model diagnostics*, *differential diagnostics*, and *exclusion diagnostics*.

Error-model diagnostics is the most straightforward form of diagnostics employing error models. It can be used whenever a system provides an output that is suspected to be erroneous. Instead of just assessing the inputs and outputs (given the assumptions of the normative system model) to identify this error, the error model (assuming that only one error is believed to potentially apply given the output) is employed to assess the presence of this error by assessing the entire system in order to provide a diagnostic evaluation. This kind of diagnostics, while being more detailed in assessment, does not lead to different conclusions than normative-model diagnostics; it is merely a more precise way to come to the same conclusion

meeting psychometric standards such as reliability and validity. On some level, these are all practices. We do things with category members. We interact with them and based on that interaction we learn how to think about (or use) the category" (ibid., pp. 222–223). The same goes for diagnostics of systems in general, so that in the struggle to do justice to a long list of practical commitments like this, it seems plausible that pragmatic considerations may lead to a categorisation of ontologically different system errors under a common error label to do justice to its commitments. For my presentation of diagnostic modelling I will set this complex discussion aside, and just assume the modeller to be interested in a maximally differential classification system.

and apply a diagnostic label from the modeller's error taxonomy. It therefore provides no diagnostic advantage over the simpler normative model diagnostic process, which is likely to make it a rarely used approach for diagnostic modellers in practice.[16]

Differential diagnostics takes place if the system's output suggests to the modeller that there is more than one error model that might match the system that comes to produce a certain erroneous output. In this case, the modeller takes the error models potentially applying to the system given the recognised erroneous output and compares them to the prepared description of the erroneous system before making a comparative judgement as to which error model applies best (above a specified threshold of good fit) for this system and selecting this model. An error-diagnostic label from an error taxonomy that is associated with the chosen model is selected and applied to the system to diagnose its error.

Exclusion diagnostics takes place if there is more than one way in which a system may be bringing about a certain error, but we do not have an error model for each of these ways. In other words, the modeller works with a knowingly incomplete set of error models for a given erroneous output. This diagnostic process then starts out in the fashion of error mode-based diagnostics or differential diagnostics, but it contains the explicit possibility that none of the error models compared may match the diagnosed system. In such case -- i.e., if one or more error models were assessed without finding them to apply to the system -- the error label assigned to the system will lead to the classification of the error by exclusion diagnostics. An error with a label identified by this way of diagnosing is more information than if it had been classified based only on the inputs and outputs of the system, but also less information than if it had been identified by the successful application of an error model. The diagnostic process has provided information that leads to a partial negative identification of the error by excluding things that might be responsible for it.

In this section I have described the process of model-based diagnostics as the use of normative models and error models for the purpose of system error diagnostics. I have also mapped out different types of error-model diagnostics. We will go into more detail on these types of diagnostics in the next chapter, where I argue that psychiatric diagnostics understood as modelling implements this approach and therefore is a kind of diagnostic modelling.

---

16 If this seems hard to grasp, imagine a case from medicine. A patient presents with an illness for which we have a symptom-based method of assessment, but we could also do more detailed biological testing. However, based on our best understanding of clinical conditions and what the tests we have at our disposal can detect, the only thing that the test could detect is the same thing that we can identify based on the symptoms. In this case, it would be a waste of effort to engage in a more detailed biomedical examination.

Figure 8: The four types of model-based diagnostics discussed so far. Arrows indicate the presuppositional relationship between the different types of modelling. For differential or exclusion diagnostics to be possible, the presuppositions for error-model diagnostics must be met. For error-model diagnostics to be possible, the presuppositions for normative-model diagnostics must be met.

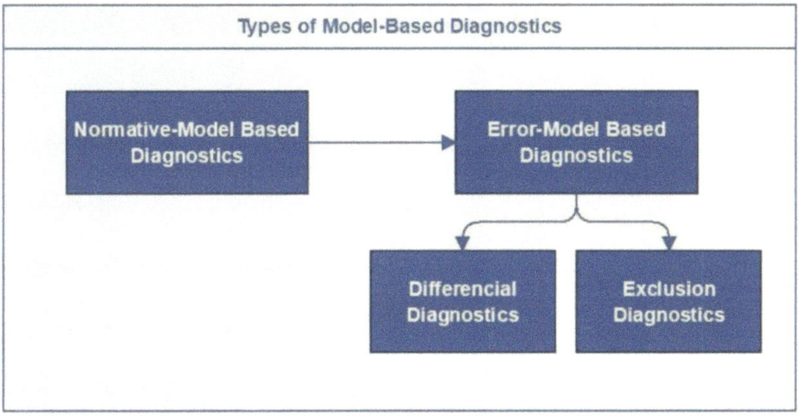

Up to this point, I have provided a stepwise description of modelling. I described modelling in general, I introduced qualitative modelling, and I presented a specific modelling procedure, namely diagnostic modelling. If we bring these elements together, the result is a qualitative modelling procedure for diagnostic purposes: *qualitative diagnostic modelling*. However, if I stopped here and moved directly to the next chapter to show that this kind of modelling maps onto psychiatric diagnostic reasoning, I would fail to answer the Methodological Question. All that this would achieve would be to provide a plausible reconstruction of the method used in psychiatric diagnostics. But this is only one of the three central aspects of the Methodological Question, as discussed in the Introduction.

What is also needed is an understanding of the rationale behind this method of modelling and an idea of how this method provides justification for its conclusions. Only if these two questions are addressed too can a demonstration that diagnostic modelling as presented here maps onto psychiatric diagnostics provide a full answer to all aspects of the Methodological Question. To ensure that my mapping attempt in the next chapter provides this full answer, I will therefore address these two aspects. Thus, when I demonstrate in the next chapter that psychiatric diagnostic reasoning should be understood as qualitative diagnostic modelling, how we should think about the theoretical rationale for following this modelling procedure and how it justifies the diagnostic conclusions will already have been developed. A successful mapping in the next chapter plus what follows in this chapter can, taken together, be considered to provide a full answer to the Methodological Question.

In the next section, I will begin this work by uncovering the inferential strategy behind diagnostic modelling, and therefore the rationale underlying this approach.

2.4 The Inferential Strategy of Model-Based Diagnostics

In this section I will explore the strategy of model-based diagnostics. When I refer to the inferential strategy of model-based diagnostics, I mean the common inferential process present in all instances of diagnostic modelling ensuring its truth-conduciveness. In other words, I mean the inferential process at work in diagnostic modelling that can reasonably be assumed to reliably led to the increase of true and the decrease of false beliefs resulting from it.[17] Understanding this strategy will, as mentioned at the end of the previous section, help us to understand why diagnostic modelling is set up the way it is, and thus to answer another aspect of the Methodological Question. The strategy at work in diagnostic modelling is what I will call it the *constitutive-indicator* strategy.

The label *constitutive-indicator strategy* derives from the idea that diagnostic modelling using normative and error models employs models that are constitutive in nature, and that differences between the system's actual behaviour and its expected operations are taken as indicators to apply certain diagnostic labels to the system. To spell out this idea and its implications, I will proceed as follows. First (2.4.1), I will discuss what it takes to be an indicator and why models in diagnostic modelling serve their epistemic purpose via being an indicator. Second (2.4.2), I will discuss what it means for a model to be a constitutive model, and how diagnostic models may be understood as constitutive models. Third (2.4.3), I will discuss the inferential patterns (inference to the best explanation, apophatic inference, inference to unintelligibility) that are realised in diagnostic modelling. Finally (2.4.4), in light of my previous discussion of the inferential strategy of model-based diagnostics, I

---

17 I focus on truth-conduciveness here, since it is usually considered the highest-ranking epistemic goal that should be supported by an epistemic practice. It is the "Epistemic Gold Standard", as Schurz (2011) puts it and as many other epistemologists also believe (e.g., Goldman, 1986, 1999; Bishop and Trout, 2005; Leplin, 2009, Ch. 2; Schurz, 2009). This position is not universal, however. Elgin (2017), for example, argued that the chief epistemic desideratum, especially in science, is *understanding* rather than truth, and that epistemic practices in science relying heavily on modelling are not aimed primarily at establishing truth in the first place. Going deeper into this discussion is beyond the scope of this project, but there have been several works that in my view convincingly refute Elgin's approach by providing alternatives to her understanding of the epistemic role of idealisation of science (Sullivan and Khalifa, 2019), unpacking the relationship between understanding and truth as epistemic desiderata, and defending truth-aptness as an epistemic priority including in science (Warenski, 2021).

will elaborate on the justificatory status of conclusions reached by a diagnostic modelling process.

2.4.1 Indication and Indicative Modelling

A widely accepted analysis of indication that I will adopt was provided by Dretske (1981). According to this analysis, an event of type *E* indicates a situation *S* to obtain if and only if the probability of *S* given that *E* occurs is 1, given some "channel conditions" under which these relationships are reliable. However, indication appears to be able to take place not only as an all-or-nothing affair; it can also come in degrees. An indicator may therefore be more or less reliable. To put it in terms of probability, some *E* may be an indicator for *S* with a probability anywhere between 0.5 and 1 [$P(S|E)>0.5$]. Different events, say *E1* and *E2*, might be better or worse indicators, depending on how reliably they indicate *S*. It seems necessary that some *E* must occur with a probability larger than 0.5 to be considered as an indicator at all. Otherwise, the "indicator" would not predict the absence or presence of a condition better than chance. You might as well flip a coin.

One feature of this understanding of indication is that the exact nature of the relationship, including the direction of the relationship, between *E* and *S* in which *E* is an indicator of *S* is undetermined. Concerning the possible nature of an indicator relationship, Dretske (1981) already noted that indicators may fulfil their role based on a causal relationship as well as a purely correlational relation.

The classic example to illustrate causal cases of indication would be the case that smoke indicates fire in certain channel conditions, since given these conditions, fire is usually the cause of smoke (fire smoke). However, this causal relationship does not enable indication only by looking at the later segment of the causal chain (smoke) to indicate the earlier one (fire); the relationship can also be used the other way around. Based on this relationship, we can also take fire as an indicator for smoke, treating the earlier segment (fire) as an indicator of the later one (smoke).[18]

18 The idea that indication may work up and down causal chains as well as via correlation is of course no philosophical achievement but has for many decades been a core topic in science interested in measurement. In psychology, for example, the terms *effect indicator* and *reflective indicator* have been around since Spearman (1904) first used factor analysis to measure general intelligence. He assumed that intelligence was the cause of the indicators he used to measure it, since changes in intelligence should lead to changes in the measured manifest indicators, but not the other way around. Today, most analysis using classical test theory, item response theory, or structural equation modelling shares this assumption. On the other hand, in psychology and social sciences we also find the term *causal indicator*, introduced by Blalock (1964), used to refer to manifest variables that can serve as indicators for the expression of a latent variable, which at the same time is theorised to be caused by this manifest variable. An example would be to consider the latent variable "life stress" to be reliably indicated by a manifest variable that would typically be interpreted to be causally

An example illustrating a correlational case would involve my dog, who rarely barks on any occasion other than when my doorbell rings, but almost always barks when that happens. Since my doorbell usually only rings when someone is standing in front of my door ringing it, the barking of my dog is a good indicator for someone standing in front of my door, but people standing in front of my door are not the direct cause of my dog's barking. People could stand there and not ring the doorbell, for example. But thanks to the channel condition that standing at my door usually goes together with it, the barking nonetheless correlates well with someone standing there. Regarding the variation in direction, in the case of a causal relationship the indicator may cause what it indicates, it may be caused by what it indicates, or it may just co-occur with it. Given this understanding of indication, let me finally say something about models.

A model might be thought of as an indicator of a state of affairs under two conditions. First, if a positive outcome of a model/world comparison using a specific model is usually reliably correlated with a given state of affairs in the system in a certain context. Under these conditions, the model's successful application can be an indicator for the relevant state of affairs. And second, if the inapplicability of a model in the context of a model/world comparison reliably correlates with a certain state of affairs, given certain background conditions. In this case, inapplicability can be an indicator of the relevant state of affairs. Let us now apply this basic idea of two forms of indication via models to see how it fits with the uses of normative and error models we encountered earlier.

As we have established, normative models and error models are the basic tools of diagnostic modelling. The attempt to apply a normative model to a system is used to indicate an error in a system, which is the case if the model is applicable to the system. Given what we said about indication, we can therefore now think of the inapplicability of a model as an indicator for the presence of a certain kind of error in the system in the context of normative-model diagnostics. Normative-model diagnostics therefore embodies one of the two ways in which models might be used as indicators: indicating a state of affairs *qua* inapplicability of a model. If we look at error models, and with them at the options of differential diagnostics and exclusion diagnostics, we find that both ways in which models might in principle be used as indicators play a role.

In the case of differential diagnostics, the modeller will consider which error model from their repertoire best applies to the system and take the one that is applicable as an indicator for a certain kind of error whose label is associated with the error model that was chosen. In this context, the applicability of a model is taken as

---

responsible for its presence, such as job loss, severe illness, or losing a loved one (Boolen and Davis 2009). For an insightful methodological debate on two conceptualisations of causally supported indication as well as covariance-based indication, see Bollen and Bauldry (2011).

an indicator for a state of affairs. For exclusion diagnostics, on the other hand, the same comparison procedure between the real-world system and the error models is pursued, but with the assumed possibility that none of the models may apply. The insight that no error model can be found to apply to the system so that no diagnostic label associated with one of the error models can be selected in these circumstances leads to an error label reserved for such an overall negative outcome being applied to the system. In these instances, it is the failure to apply one (or several models) that is taken to be an indicator to ascribe a diagnostic label to a certain state of affairs.

We have now discussed the indicator portion of the constitutive indicator strategy for diagnostic modelling, and thus established what indication is and how to think of models in general, and specifically normative and error models, as indicators. Now let us turn to the other portion of the strategy constitution. In the following discussion of this topic, I will show that the models used for indicational purposes by diagnostic modelling make constitutive assumptions about the phenomena that they are meant to indicate. As I will show, the models' constitutive nature is important because the resulting explanatory relationship between the models and phenomena ensures a reliable relationship that makes it plausible that they reliably indicate the targeted phenomena in the first place.

2.4.2 Constitution and Constitutive Modelling

The technical term constitution was already used in the last chapter with reference to Tyler Burge (2010, p. xv). Burge thinks of constitution in terms of constitution conditions – that is, the necessary and sufficient conditions for something to be what it is. In this context, a different understanding of constitution shall be considered. The understanding of constitution relevant here derives from debates about constitutive explanations rather than identity conditions. What is the difference? While constitution in Burge's sense concerns the conditions for something being what it is, addressing constitution in terms of a constitutive explanation aims to provide an explanation of a system's causal capacities, by giving an understanding of these capacities through reference to the system's parts and organisation (Ylikoski, 2013).[19] To understand a constitutive model, in other words a model that can be assumed to do explanatory work in terms of a constitutive explanation, a better un-

19 These two senses of "constitutive" differ regarding the *explananda* they can target, they involve different interpretations of what it is to spell out something's constituents, and they use different explanantia to provide answers to constitutive questions.
The potential difference regarding the explanatory target is quickly spotted. While constitutive explanations target causal capacities, Burge's account of constitution is not limited in scope in this way. Burge can ask what constitutes any phenomenon, including causal capacities but not limited to them. Burge could ask constitutive questions about the last fish in the ocean ("What is it to be the last fish in the ocean?") or Latin dance ("What is a Latin dance?")

derstanding of these constitutive explanations themselves is necessary. Therefore, I will proceed to flesh out the idea of constitutive explanation.[20]

---

that are excluded from being addressed by a constitutive explanation approach, since there are apparently no last fish in the ocean and being a Latin dance is not a causal capacity.
To see the extent to which the two approaches differ in their understanding of what it is to seek out something's constituents, we can imagine a constitutive question that both may address and see how it would be interpreted by them. For this purpose, think of the question "What makes the glass fragile?". Taking this question in Burge's sense would mean asking "*what does it mean to be fragile?*" and "*are sufficient conditions for fragility met by the glass?*". Interpreting the question in the sense of a constitutive explanation would mean asking "*why, in terms of its physical parts and their organisation, is the glass fragile?*". The first interpretation of the question is about spelling out what about the glass makes it true that it is fragile given the identity conditions of fragility, while the second question is about how the parts of the glass and their organisation are responsible for it being fragile. One question is about what it is to be fragile for the glass and the other is about how it happens that the glass has the feature of fragility. Note that on both understandings, the constitutive questions are vastly different from causal questions. A causal question with regard to the fragility of the glas would ask "*how did the glass become fragile?*" or "*why did the glass break?*". Answers to this question would tell an aetiological story about how the glass broke or how it came to have the disposition to break. But the difference between the two non-causal but constitutive ways to take the question about the fragility of the glass translates to relevant differences between these two in terms of the explanantia that can be called upon given each understanding.
These approach-dependent differences in answering constitutive questions arise partly because identity conditions may not always be exhausted by statements about a system's parts and organisation. This may be true in cases in which the question itself does not concern a causal capacity. An example can be taken from Burge (2010, p. 379). In his view, one of the constitutive conditions of perception is that every perceptual state must have a veridicality condition. Having veridicality conditions, however, is not a matter of the parts and organisation of a system, but a condition that is an intellectual normative/epistemic property of perceptual states. Putting forward something's constitutive identity conditions is therefore not necessarily the same as providing a constitutive explanation in terms of a system's parts and organisational features, but it makes use of other kinds of explanation – here, normative/epistemic properties. Moreover if one accepts the possibility of multiple realisation of causal capacities, for example "being sighted", a statement about the parts and organisation of any specific types of a system's parts and their organisation instantiating this capacity may fall short of providing a list of conditions that would suit a constitutive answer in Burge's sense. In such cases, a statement on the parts or the organisation would fail to point out necessary conditions required for being sighted. A burgean answer presenting the identity conditions for the causal capacity would then have to provide identity conditions other than parts and organisations of systems, which – whatever they turned out to be – would have to be something beyond scope of the explanation used by constitutive explanations.

20 The most important points about constitutive explanations and the debates surrounding them have been usefully reviewed by Ylikoski (2013), on whose efforts I rely in the following.

Constitutive explanations are non-causal, and like most non-causal kinds of explanation, constitutive explanations have received relatively little philosophical attention compared to causal explanations. Relevant early work on constitutive explanation was done by Cummins (1975, 1983), complemented more recently by Craver (2007a). Constitutive explanations, as mentioned, aim to explain the causal capacities of a system (Cummins 1975, 1983; Harré and Madden 1975). These causal capacities are understood as the dispositions of a system to bring about a certain causally influential event or occurrence, given specific triggering or enabling conditions. In absence of these enabling conditions, causal capacities are powers existing unactualised in the system. These causal capacities can be explained by the parts of the system and their organisation, by virtue of which they are present in the system. Providing such an explanation, however, is not providing a causal explanation because it does not provide a causal story of *why* the system is doing what it is doing by spelling out the causal aetiology of its behaviour. Instead, the explanation accounts for *how* the system's components and their organisation instantiate the causal capacity to make this behaviour happen. As Cummins (2000, p. 122) lucidly sums it up: "The constitutive questions abstract away from the behavior and orchestrated activities of the parts, and ask how the system has a capacity for this kind of behavior."

This being the basic idea of a constitutive explanation, there are three further important peculiarities that I will also discuss in elaborating how diagnostic modelling is to be understood as constitutive. First, constitutive explanations are provided in what has been called a constitutive field; second, the scope of constitutive explanations usually entails single dispositional features; and third, constitutive factors used for an explanation may be used at various levels of description and grain. Let me start out with the first point, concerning the constitutive field. Now that we have an understanding of what constitutive explanations are, let me supplement this understanding with some ideas about *how* these explanations explain.

To understand how constitutive explanations explain, Ylikoski (2013) suggests combining two general approaches to explanation from philosophy of science: the *contrastive question approach* (e.g., Garfinkel, 1981; Hesslow, 1983; Lipton, 1991; Ylikoski, 2007; Craver, 2007b) and the *difference-maker* account (e.g. Mackie, 1974; Woodward, 2003; Waters, 2007; Strevens, 2008). The idea is that constitutive explanations have explanatory power because they treat their explanatory questions as *contrastive questions* that they answer by identifying the *difference-makers* responsible for the factual differences pointed out by the contrastive questions. What this means needs some explanation.

According to the contrastive question approach, the epistemic value of explanations is that they tell us why some fact rather than some other exclusive alternative fact (or group of facts) holds true. The exclusive alternative facts considered for this purpose are called the contrast class. When we ask an explanatory question, we do not always put forward a contrast class explicitly; rather, the contrast class is often implicit in the background. However, awareness of the assumed contrast class is central to clarity about what exactly it is that ought to be explained by an answer to an explanatory question. If one has such awareness, rather than only asking why X ϕ´s, the question is why X ϕ´s rather than X ψ´s, where ϕ and ψ are exclusive alternative facts about the subject X. The contrastive question approach assumes that all explanatory questions can be understood along these lines. Whether one believes this or not, let us for now follow Ylikoski in his claim that this understanding is at least useful for grasping constitutive explanations.

As an example, consider the explanatory question of why a certain animal species is found in the Atlantic Ocean. Although one might have an intuitive understanding what kind of explanation this question is asking for, it is actually ambiguous. It does not tell us what about the fact that this species is found in the Atlantic should be explained. It could be a) that it lives in this ocean rather than not living in it; b) that it lives in this ocean but not in *any* another ocean, or that it lives in this ocean but not in a specific other ocean; c) that it lives in this ocean but not on land as well; (and so on). Alternatively, the question could aim to address all these contrasts at the same time. Depending on which of these contrastive facts are considered to be part of the contrast class assumed for the question, the answer to the question will look very different. An explanation of why a species lives in in the Atlantic Ocean at all rather than not living there will obviously look different from explaining why it lives there but not also somewhere on dry land. Of course, in principle one can include every possible alternative fact in the contrast class. This would make matters incredibly complex, however, and turn a question that needs one answer into a question addressing multiple things that need to be answered separately at the same time. If a question is thought about clearly, it usually ends up having only one contrastive fact; indeed, a question may need dividing into a set of questions if it had more than one item in its contrast class before.[21]

Applied to constitutive explanation, the contrastive question framework considers the question asked to be why a system has a causal capacity opposed to alternative exclusive facts. Again, a precise explanatory question will have a specific difference that is intended to be explained, set by the chosen contrast class. If, for example,

21 As Ylikoski (2013) points out, we can also use the approach to reference classes to determine the difference between causal and constitutive questions: "Thus the contrastive thesis is not a claim about what people have in mind when they put forward an explanation-seeking question, but what they should have in mind (Ylikoski, 2007). Quite often the original scientific research question, when articulated in contrastive terms, turns out to be a whole set of related contrastive questions. This is a good thing: smaller questions are something we can actually hope to answer by means of scientific enquiry. And of course, nothing prevents one from asking both causal and constitutive questions – and questions driving scientific research are often such – but the contribution of the contrastive idea is to make it possible to analytically distinguish questions that require separate answers" (ibid., p. 123).

what should be explained constitutively is the fragility of a glass, this could be spelled out as the contrastive question for what makes a glass fragile, as opposed to the assumed contrast class. This contrast class may contain the property of robustness, so that a constitutive explanation has to answer which constitutive factors make the glass fragile rather than robust. Alternatively, the contrast class might contain the disposition to liquify under force. Whatever the contrast class looks like, it directs the explanatory effort to what should be constitutively explained about a present causal capacity.

The second component of how contrastive explanations explain is the difference-maker account. While the contrastive question account provides a better grasp on what the exact target of an explanation (its *explanandum*) is, the difference-maker account addresses what the explanatory components (the *explanans*) should be. The idea of this account is that a targeted fact or state of affairs is explained by pointing out what is responsible for this state or fact (as opposed to alternative states or facts) obtaining. The explanation thus identifies counterfactual dependencies, claiming that if the *explanans* had differed, the *explanandum* would have been different too. Combined with the contrastive question account, the task is then more precisely to identified counterfactual dependencies on responsible difference-makers for the targeted fact hold true, as opposed to the alternative facts contained in the contrast class. As Ylikoski (2013, p. 291) puts it:

> The idea of the explanans as the difference-maker is a natural partner of the idea of contrastive explanandum. Together they provide a powerful heuristic of scientific research. First, you create, find, or imagine the difference to be explained, and then you proceed to find the differences between the cases. Then you test whether these candidates can really make the difference, by testing whether they can bring about the difference to be explained.

In causal explanations, this would mean providing an understanding of the counterfactual dependency of an event Y on a previous event X, such that if X had not happened Y would not have occurred, given certain background conditions (the relevant constitutive field). Here X is the difference-maker causally explaining the presence of Y (Woodward 1984, 2003; Ylikoski and Kuorikoski, 2010).

The difference-maker approach also matches with constitutive explanation. Considered along these lines, providing the constitutive explanation means providing the conditions for the fact that a certain causal capacity rather than a chosen contrasting capacity or the absence of this capacity is realised in the system, by providing a statement about the parts and their organisation in a system (given certain constitutive field conditions) on whose presence the fact that the capacity in question is in fact present counterfactually depends. Constitutive explanations explain by providing such counterfactual conditions as difference-makers.

In the preceding paragraphs I have taken some time to introduce constitutive explanations in their structure and explanatory capacity. Now let me show how this fits together with modelling in general and diagnostic modelling in particular, in order to spell out the constitutive aspect of the constitutive-indicator strategy pursued by diagnostic modelling.

In general, for a model to qualify as a constitutive model, the model will be required to entertain the constitutive factors of a phenomenon in the model. More precisely, these constitutive factors, which make up the *explanans* of the constitutive explanation of a phenomenon, must be covered within the set cope of the model by being assigned to parts of the model structure, with the purpose of making the model a model of what would be targeted as an *explanandum* by the corresponding constitutive explanation. A model build based on a constitutive explanation of a phenomenon would therefore take the explanation as a background theory for setting up the model structure, deriving its plausibility as an adequate model of the phenomenon of interest from the plausibility of the background theory (the constitutive explanation) it is capitalising on. If the model were then used in a diagnostic manner – that is, to indicate the presence or absence of a constitutively modelled phenomenon in the targeted system as result of a model/world comparison – then the overall plausibility that this model could be used for such indicative purposes would rely on the quality of the constitutive explanation. If a constitutive explanation is assumed to capture the relevant constitutive factors of the disposition targeted by the model, it can be assumed that their presence indicates the model's target to be present, and thus that the model is a valid indicator. By using a constitutive model along these lines as an indicator, it also provides an explanation. More precisely, it provides a constitutive explanation in terms of difference-making. By pointing out the presence or absence of relevant constitutive factors that make the difference between the presence or absence of the disposition, the constitutive model tested in the model/world comparison provides a constitutive explanation by accepting or rejecting the compared model as adequate, and thus explains by saying that the factors that make the difference between absence and presence of the disposition apply or not.

Equipped with an overview of constitutive modelling and its indicative use in general, let us now turn to its use in the specific cases of diagnostic modelling. For this purpose, I will again speak of normative models as well as error models.

As established earlier (2.3), normative models give an idea of the diagnosed system in terms of its disposition to realise a certain causal capacity (output) given certain triggering conditions (input). The identification of a certain type of error in normative-model diagnostic reasoning takes place based on the type of abnormality that shows up in the comparison between the normative model and the actual behaviour of the system. According to the way it behaves differently from the normal input-output behaviour, the system can then be classified as presenting a certain type of error. In the assumptions regarding the normative model, we therefore find a set of nested assumptions about what constitutes a normal system behaviour given an array of different inputs. In other words, the disposition to behave normally given certain triggering or input conditions is constituted by providing a certain output in response. Thus, the normative model is an amalgamation of a set of assumptions about what constitutes normal behaviour for the system in terms of inputs and outputs. The inapplicability of the model, given certain inputs, is then considered an indicator of the abnormal functioning of the system in one of the functions covered by it. The inapplicability thereby explains the assumption of the presence of the error constitutively by the absence of the constitutive factors that would render the system free form error.

In the case of error models, they too are structures geared towards matching up with constituents, but these models differ from normative models in two relevant regards. First, they are targeting constituents of the system's disposition that are considered to be responsible for an error occurring in the system, rather than those responsible for normal functioning. By identifying the presence of the constituents of a specific error, constitutive error models thereby explain the presence of the error by identifying the relevant difference-making factors responsible for it. And second, every error model is intended to target one specific instance of error to diagnose one specific error, and is thus informed by one constitutive explanation rather than being an amalgam of multiple explanations as in the case of a normative model. Their use as an indicator then works in differential diagnostics by testing their applicability in model-world compression with the erroneous system. If the presence of the specific set of constituents assumed in the error model can be found in the real world system this justifies the model's acceptance and thus justifies to apply the diagnostic label that the error the model is considered to be a model of. In the case of exclusion diagnostics, the inapplicability of any of these models – and therefore the inability to identify the relevant constituents for any condition that may be ascribed based on the application of error models – may be marked by a corresponding diagnostic labelling of the present error in the system that could not be modelled more specifically by an error model. Thus error models used in exclusion diagnostics explain by noting the absence of any set of relevant difference-makers that would indicate the presence of a certain error entity.[22]

---

22 In line with my remarks in footnote 18, where the discussed indication may rest on a causal or a correlational relationship, and noting how these options are mirrored in contemporary methodological debates in psychology, I want to point out something similar for constitutive indicators. While the idea of a constitutive indicator has to my knowledge not been discussed in detail in philosophy, it is present in the psychological research literature. There, the kind of indicator I am calling *constitutive* is called *composite*. Similarly to my discussion, Bollen and Bauldry (2011, p. 6) introduce them by saying that "Composite indicators are weighted elements that form a composite variable for which there is no disturbance term. That is,

Now that I have discussed the constitution aspect of modelling following the constitutive indicator strategy, I will now analyse what inferential patters are facilitated by this strategy. Doing so will be especially relevant to the last section of this chapter, since I will address the question of how qualitative diagnostic modelling justifies its outcomes by following the constitutive indicator strategy.

2.4.3 Inferential Patterns of Model-Based Diagnostics

In this section, I will discuss two inferential patterns occurring in the context of model-based constitutive indicator diagnostics. The first type of inference made in model-based diagnostics is *abduction* or inference to the best explanation (IBE) (2.4.3.1).[23] This type of inference occurs when the diagnostic modeller engages in basic normative-model diagnostics and in error-model diagnostics as forms of differential diagnostics. The second type of inference is *apophatic* inference (2.4.3.2). The name derives from ἀπόφημι (from *apophēmi*, "to deny") and is an adjective meaning "involving knowledge obtained by negation" (Harper, 2023). This type of inference, or rather (as I will discuss later) one of its specific instances, occurs if the normative model can be applied to the system and an error occurs, but as a result of the diagnostic evaluation none of the diagnostic error models for this error can be applied – in other words, in exclusion diagnostics.

the composite variable is an exact linear combination of the composite indicator variables. But beyond having no disturbance, the composite indicator coefficients are not structural or causal coefficients. Rather their coefficients are weights to apply to form the composite variable that is made up of them." Furthermore, they also bring up another point that was introduced in my discussion of constitutive explanations – namely, that constitutive explanations can address various aspects of the system that are relevant to realising a power of interest in the system on various level of description, a quality that plausibly carries over to constitutive models as well. As they put it, in discussion of social variables (e.g., character traits as variables): "Composite indicators do not necessarily have conceptual unity, but can be an arbitrary combination of variables" (ibid.).

23 Since my aim is to bring the theoretical considerations of this chapter to bear on psychiatric diagnostics, let me point out here that I am aware that the claim that IBE plays a role in medical diagnostics is not a new idea (e.g., Lipton, 1991; Console and Torasso, 1991; Gabbay and Woods, 2005; Aliseda and Leonides, 2013; Johnson, 2019; Stanley and Nyrup, 2020) and that it has also been raised with regard to psychiatry (e.g., Reznek, 1998; Vertue et al., 2008). What is unique about my account and what makes it preferable will be outlined in the final chapter.

Table 2: Types of inference (left) matched with inferential practices (right)

Type of inference	Corresponding diagnostic practice
Inferences to the best explanation (or abduction)	Normative-model diagnostics Error-model differential diagnostics
Apophatic inferences	Error-model exclusion diagnostics

2.4.3.1 Abduction or inference to the best explanation

Abduction as a reasoning pattern was introduced by Peirce (1878, 1903), in contrast to the two other widely discussed patterns of reasoning, induction and deduction. Deductions are *non-ampliative* (they do not add to what is already known) and *certain* (their conclusions must be true if their premises are true). Both inductions and abductions are *ampliative* and *uncertain*, which means that if true they extend our knowledge of the world, but that in contrast to deductions, their truth is not guaranteed by the truth of their premises. In inductions, properties or regularities are transferred from past events to the future, or from the observed to the unobserved. The difference between abduction and induction regards their target. While inductions aim to make inferences about future or unobserved events, abductions aim to infer something about the unobserved causes or explanatory reasons for an observed event (Aliseda, 2006). Since the distinctions amongst these three types of inferences were introduced, philosophers have recognised that abduction itself is not a single pattern of reasoning but consists of a collection of patterns of inference. One particularly valuable attempt to defend and systematise the various patterns of abduction or inference to the best explanation (IBE) is provided by Schurz (2008).

Following Schurz (2008), "the crucial function of a pattern of abduction or IBE consists in its function as a *search strategy* which leads us, for a given kind of *scenario*, in a reasonable time to a most promising explanatory conjecture test which is then subject to further test" (ibid., p. 205). This function can be fulfilled in different ways following different patterns of abductive reasoning. These different patters fall into two broad classes: *selective abductions* and *creative abductions*. In selective abduction, the task is to make a choice between competing alternatives that might explain the features the target phenomenon, while in creative abduction the task of the reasoner is to come up with a new explanation given the *explanandum* and potential constraints deriving from further circumstantial knowledge (ibid). Given my description of the inferential strategy of model-based diagnostics, I will concentrate on selective abductions as the one most plausibly at work in this form of diagnostic practice. More specifically, I will concentrate on a class of abductions that Schurz calls *factual abductions*, and its subtype of *observable-fact abduction*.[24]

Factual abduction is the classic and most widely discussed form of abduction, introduced by the young Peirce (1878) himself (calling it *hypothesis*), before he generalised his understanding of abduction (Pierce, 1903) along the lines presented earlier. As Schurz (2008, pp. 207–208) puts it:

> In factual abductions, both the evidence to be explained and the abduced hypothesis are *singular facts*. Factual abductions are always *driven* by known implicational laws going from causes to effects, and the abduced hypotheses are found by backward reasoning, inverse to the direction of the lawlike implications. (…) It has the following structure (the double line === always indicates that the inference is uncertain and preliminary):
>
> (FA): *Known Law:* If Cx, then Ex
> *Known Evidence:* Ea has occurred
>
> ===========================
>
> *Abduced Conjecture:* Ca could be the reason.

Observable-fact abduction is a sub-pattern of factual abduction. As Schurz argues, it occurs if there is a follow-up test procedure for the abduced conjecture such that "the follow-up test-procedure consists in the attempt to gain direct evidence for the abduced conjecture" (ibid., 207). Schurz offers the example of a murder investigation: "In the example of a murderer case, such direct evidence would be given, for example, by a confession of the putative murderer to have committed the crime" (ibid.).[25] Let us now see how the two inferential (sub)patterns of abduction, *factual abduction* and *observable-fact abduction*, apply to diagnostic modelling.

First, factual abduction applies to simple normative-model diagnostics. To recap, in this kind of diagnostic modelling, the modeller first recognises that the system produces an output that does not seem to be in line with its expected normal be-

24 Other forms of abduction irrelevant to my purposes but discussed elsewhere include law abduction, second- order existential abduction and its subtypes (micro-part abduction, analogical abduction, hypothetical cause abduction, speculative abduction), common-cause abduction and its subtypes (strict common-cause abduction, statistical factor analysis, abduction to reality), and theoretical-model abduction. If you are interested in these, I suggest you consult Schurz's excellent (2008) paper.

25 Philosophers such as Fumerton (1980, p. 592 f.), have claimed that abduction could be reduced to induction. While I am not able to discuss this claim here in detail, please see Schurz (2008, p. 207 f.) for a counterargument.

haviour. Then, to determine whether the system showing the *prima facie* erroneous output really is in error, the modeller has to ensure that their normative model really indicates a deviation of the system from normal behaviour in this situation. If this is the case, the system's behaviour can be classified as presenting an error. The inference taking place can then be mapped onto factual abduction. The modeller's background assumption is that a certain a kind of erroneous output of the system (Ex) usually occurs as a consequence of some (not further specified) alteration of the system (Cx), such that if there is a relevant sort of alteration (Cx) then the error occurs (Ex). That the error has occurred in the system (Ea) then justifies the inference that some error in the system is present (i.e., that Ca could be the reason).

The more specific subtype of IBE, *observable-fact abduction,* occurs in the case of error-model differential diagnostics. In this case, diagnostic inference is not based solely on a system's deviation from its behaviour as predicted by the normative model, suggesting some constitutive alterations in the system. In addition, and more specifically, error-model differential diagnostics takes place in the form of evaluation of the specific changes occurring in the system against specific error models. These models represent potential alterations of the system that may constitute the system's disposition to produce the error. Using these models can serve the diagnostic process in terms of differential diagnostics in two ways.

The first way for error models to serve differential diagnostics occurs if only one error model is known that should be applicable to the system if a certain error occurs. In this case, the error model may apply, and if so, the error model further supports the diagnosis provided based on the initial normative-model diagnosis, by showing that the specific setup of the system that is known to potentially bring about the error can indeed be found in the system. Alternatively the error model does not apply; therefore the diagnostic conclusion will be that the error initially identified with the aid of the normative model is present, but that it is an instance of the error not covered by the diagnostic understanding provided by the error model. This scenario turns the process into exclusion diagnostics, which will discussed in more detail below in connection with apophatic inferences.

The second way occurs if more than one potential error model exists that might match the system to explain the occurrence of the error beyond what could be said based on the normative model, and if indeed one of these models applies. In this case, the error found in the system can be identified as a specific instance of the initial error and can therefore be classified by a more specific diagnostic label. Again, it might turn out that no error model applies, which would, as before, lead to exclusion diagnostics, to be discussed in the context of apophatic inferences.

If we stick to the cases in which the modeller is successful in their attempt to apply an error model to the system, observable-fact abduction takes place. In this case, beyond the previously illustrated step of normative-model diagnostics and its factual abduction, an additional round of abduction following the same schema takes place. This time, the modelling process does not take an erroneous output of the system as evidence; instead, it takes more specific features of the system as constitutive, whose application is supposed to indicate a more specific error than the rather abstract error attribution based on a normative model. This act of *observable-fact abduction*, looking for specific evidence to support a diagnostic claim, may thereby test the error model that would support the initial diagnosis of a normative-model based diagnostic conclusion. Alternatively, if the modeller's understanding of the system is more differentiated, such that multiple types of error might stand behind the error that would be recognised based solely on normative-model diagnostics, a differential diagnostic process would take place in which multiple error models would be applied to the system to make an observable fact abduction to the more specific error type they suggest. As mentioned earlier, if no model applies, the modeller may instead end up in an exclusion diagnostic process, involving apophatic inference, which I will discuss next.

2.4.3.2 Apophatic Inferences

Next in line are cases of what I will call *apophatic* inferences. Apophatic (from ἀπόφασις, to deny) inferences are not an inferential pattern in themselves, but they are instantiations of the commonly discussed inferential patterns (induction, deduction, and abduction) that draw *negative conclusions*. In philosophy, apophatic inferences have been discussed since Plato, and became especially prominent in Middle Platonism and Neoplatonism, via the still existing branch of theology called negative theology (Westerkamp, 2006).[26] For the analytic tradition, the idea of attaining knowledge by negative conclusions is also a familiar one, thanks to Popper's (1935) emphasis on falsification in the critical-rationalist approach. However, in recent years apophatic inferences have attracted attention mainly outside of the analytic tradition.[27]

Before I come to how apophatic inferences occur in model-based diagnostics, I shall begin by saying more about the nature of negative conclusions, their truth conditions, and their informational value. First of all, I will say something about how I will handle the most distinctive feature of apophatic inferences: negation. Negation in natural language and logic is a complex topic, a comprehensive treatment of which is beyond the scope of this chapter. For my purposes here, I will focus on

26 Please note that my use of the label *apophatic* does not suggest that there is a full match between the methodology of negative theology and the types of inference I describe here. I chose the label because I see a broad resemblance in the type of approach – namely, a pattern of inferences trying to arrive at an ultimate statement about a target by means of negative ascriptions.

27 Indeed it seems that the most recent debates in philosophical circles that have tried to actualise the idea of the *via negativa* have taken place among theologists and philosophers sympathetic to poststructuralist philosophy (e.g., Derrida, 1995; Ferretter, 2001; Rubenstein, 2003).

negations understood as indicative-mode declarations of negative predications, as originally discussed by Aristotle (De Interpretatione, 17a25). In other words, I will adopt the view that negations are statements consisting of a subject and a denied predicate applied to a subject that together form a proposition: C does not apply to a.[28] With this clarified, let me next discuss what kind of information we can gain from negative conclusions.

Prima facie, negative statements do not seem to correspond to specific facts that would serve as truthmakers of the negative statement in question. Rather, the informative value of such statements seems to lie within the information about the absence of the truthmakers of a state of affairs denied by the statement.[29] Therefore the informational content can be derived almost trivially from the negation itself: If the negative statement is true, it is not possible that any set of minimally sufficient facts from the set of all necessary and sufficient facts that would be truthmakers of the positive formulation of the negative statement hold true.

---

28 Why did I choose the Aristotelian model of negation considering negations to be part of propositions (C does not apply to *a*), rather than the model of negation from Fregean logic that considers negations to be denials of propositions ("it is not the case that *p(p=Ca)*")? The reason lies with the scope of the negation that makes each of these negations true. A negation expressed according to Fregean logic would be true under two circumstances: first if C does not apply to *a*, and second if there is no *a*. While the negation according to the Aristotelian logic claiming that C does not apply to *a* is true if there is an *a* and C does not apply to it, it is false if there is no subject on which the C could be predicated. Intuitively, these conditions make the Aristotelian model closer than the Fregean model to our natural language use of negation, and also closer to the use of negative statements in diagnostics. If I make a diagnostic statement that a certain error does not occur in a system, to claim that this statement is right since the system I am talking about does not exist seems strange. Rather than saying that this statement was right, it seems plausible to say that this statement is wrong or meaningless because the system I am talking about does not exist. Consequently, the Aristotelian model of negation seems more adequate to understanding negation in the context of system diagnostics. For a more in-depth discussion of Aristotle's understanding of negation and its defence against criticism from modern logicians, see Perälä (2020). For a comprehensive discussion of negation in natural language and logic in general, see, e.g., Horn and Wansing (2020).

29 In this point I basically side with Lewis (2001) approach to truthmakers in that I do not think there are specific facts that are truthmakers for everything that is (or can be) true. Rather, my view is that negations are true due to the fact that in current state of affairs, facts that would be truthmakers for the affirmative equivalent of the negation do not hold. This approach helps to avoid problems occurring if one begins to look for specific facts serving as truthmakers of negations, such as the so-called Paradox of Negation that concerns the questions "If a positive statement refers or corresponds to a positive fact, to what state of affairs does a negative statement refer or correspond?" and "What in fact is a negative fact?" (Horn et al., 2020).

Taking this for granted, it appears that information attained by a negative judgement as a conclusion of an inferential process is therefore relatively limited. This apparent poverty of negative statements was already pointed out by Plato in *The Sophist* when he stressed that it is in the nature of negative judgements to suffer from a lack of specificity, as all we learn from them is what is not the case, making them in general less informative than positive judgements.[30] This, however, is not strictly true. The informativeness of negative judgements depends partly on the context of their assertion, more precisely on the space of possibilities that forms the contrast class to the negative judgement. The relation is such that the smaller this contrast class is, the larger the informative value of a negative judgement becomes. Let us look at an example. If I make the negative judgement that my grandfather is not alive, this judgement has a high informational value given that the relevant contrast class of "being alive" contains only one alternative if we apply it to people who are already born, namely "being dead". The informational value is lower if, for example, I make the judgement that my father is not a bachelor, as the relevant contrast class to "being a bachelor" contains not one but several options. The man might be a fiancé, a spouse, or a divorcé. Even less informative would be the statement that something is not green, or, even worse, that something does not weigh 15 kg, since the intuitively chosen relevant contrast classes (i.e., all other colours or all other possible weights) form larger and larger contrast classes. From this it is clear that the scope of possible alternatives seems to determine the informative value of negative judgements. The claim that negative judgements are in *general* less informative than positive judgements has to be specified by saying that they are less informative as long as there is more than one alternative exclusive state of affairs, and they become the less informative the more such alternatives exist in the relevant contrast class.

Now that the informative value of apophatic inferences has been discussed, let me come to the relevant instantiation of apophatic inferences in model-based diagnostics. They occur as deductive inferences, instantiated as *modus tollens:*

(AI – D): *Known Law:* If Cx, then Ex
Known Evidence: Not Ea

Apophatic conclusion: C does not apply to a

Let us see how this applies to model-based diagnostics. Here such apotheic judgements occur if the behaviour of a system *prima facie* suggests a certain type of error

30 See Xenakis (1959) and Lee (1972) for a detailed treatment of Plato's thoughts on the informativeness of negative statements.

and a normative model is applied to the system – that is, normative-model diagnostics has taken place – but if, as a result of closer diagnostic evaluation based on more specific diagnostic error models, none of the tested models applied.

The attempt to apply these models to find the correct diagnosis is made with each of the models considered to suggest the presence of a certain kind of error (If Cx, then Ex / If Kx, then Lx/ ...). However, if it turns out that none of the models applies (Not Ca/ Not Lx/ ...), then none of the observed errors can be diagnosed (E, L, ... does not apply to a). As a result, a finer-grained diagnostic judgement is not feasible. While the initial diagnostic evaluation *qua* abduction allows us to determine a type of error to be present, the second level of evaluation is based on more specific error models that add information about what potential instance of this error is not taking place in this system. The result is an instance of exclusion diagnostics.

So far, model-based diagnostics as introduced in the first half of this chapter has been elaborated regarding its inferential strategy and the inferential pattern at work in it. Presenting the inferential strategy has made clear the rationale for believing that this approach achieves its epistemic goals of correctly indicating and classifying errors in a system. Discussing the inferential patterns at work in model-based diagnostics has clarified what justificatory procedure is present in which aspect of model-based diagnostics. From this, I will now transition to the closely related question how we should think of the justificatory states of results obtained from a model-based diagnostic process.

2.4.4 Model-Based Diagnostics and Justification

To discuss the justification of conclusions in model-based diagnostics, I will distinguish between their internal epistemic justification (2.4.4.1) and external epistemic justification (2.4.4.2). Essentially, when I talk about the internal justification of diagnostic conclusion in model-based diagnostics, I mean the epistemic source of a justification a conclusion received within an assumed diagnostic system (a set of diagnostic models to diagnose a certain system). When I talk about external justification, I am referring to the source of justification that is outside the system insofar as it provides reason to trust the framework of a diagnostic system used for diagnostic modelling in the first place. To put it briefly: internal justification is concerned with the source of epistemic justification *within* the diagnostic procedure, while external justification deals with the diagnostic justification *of* the diagnostic procedure. Let me expand a little more on both types of justification to ensure that the difference is clear.

In internal justification, the justification enjoyed by the diagnostic conclusion within an adopted diagnostic system of model-based diagnostics is achieved by virtue of meeting the internal standard assumed by model-based diagnostics as a strategy to arrive at its diagnostic conclusions. By *internal standards* I mean the epistemic norms of diagnostic procedure that a given diagnostic system needs to follow in order to arrive at a diagnostic conclusion considered adequate within this framework. Looking at internal justification allows us to identify, for example, how conclusions within an established diagnostic system come to be deemed justified. As I will argue, the epistemic core value relevant to internal justification in diagnostic modelling is reliability.

External justification, on the other, is the justification a diagnostic conclusion enjoys by virtue of being a product of an epistemic procedure meeting the "epistemic gold standard" (Schurz, 2011) – namely, being truth-conducive.[31] Whether the epistemic procedure of diagnostic modelling meets this gold standard will not depend on the plausible internal framework used to justify its conclusions, but will rather be based on how good our reasons are for claiming that the procedure that is producing results is indeed producing correct results. In other words, the question is whether a diagnostic modelling process is following the general approach and employing a certain set of diagnostic models to diagnose a system in a way that is reliable, and of which we also have reason to believe that its outcomes track actual instances of errors in the system. What is at stake here is the validity of a given modelling procedure. As I will argue, this validity depends on the quality of constitutive explanations that are used to infer the absence or presence of certain error conditions.

Discussion of the justification of conclusions in model-based diagnostics is crucial to for allowing us to address the Methodological Question. It is crucial because in a methodology we want to understand how a method justifies. Discussing internal and external justification separately for this purpose is important to ensure that the considered method follows an internally rational route to come to conclusions that we can make comprehensive in a theory of this method (internal justification). Beyond being internally comprehensive, it is also important that we have reason to believe that a method performs well in its application to the real world and that we should trust its results, or at least that we know to what extent we can trust its results (external justification). I will begin by addressing internal justification.

2.4.4.1 Internal Justification

To address the internal justification of diagnostic conclusions, let me quickly review some aspects of the model-based approach. Diagnostic modelling follows the constitutive indicator strategy. In brief, this means that diagnostic conclusions in a given diagnostic system are drawn by testing the (in)applicability of normative and error models. The results of these comparisons are then used in different ways (normative-model diagnostics, error-model diagnostics) to indicate the absence or presence of errors. Since the occurrence of a (mis)match of a model used in the diagnostic

---

31 For a brief discussion of this standard view, see footnote 17, chapter 2.

process is taken to be an indicator, and since this indication is based on reliable correlation, our trust in the results rests on their justification *qua* the epistemic value of *reliability*. To bring an example to mind that highlights the centrality of reliability in the context of indication, think again of the example of the doorbell and the barking dog discussed in my earlier analysis of indication. What makes the barking dog a good indicator that the doorbell rang is that the dog almost always barks when the doorbell rings, and rarely barks on any other occasion. The barking is a good indicator because of its reliability. The same is true for diagnostic models: they are thought to be good indicators because of the reliable correlation of their (mis)match with the targeted system in case of the presence of the error they are intended to indicate. While we can thus say that reliability is crucial to the internal justification of diagnostic conclusions, one may expect there to be more to say about this. More specifically, one may hold the *prima facie* plausible intuition that the strength of internal justification for diagnostic conclusion in model-based diagnostics may depend on the type of inferential procedure used to produce it. Let me elaborate why one may think so.

One may think that although all inferential patterns used in model-based diagnostics rests on the justification by reliability, some of these patterns may provide better justification to conclusions than others. Should we not expect, for example, that error-model diagnostics would be better justified than normative-model diagnostics, given that, as we discussed earlier, error models assume a far more detailed understanding of specific errors that must be found present in diagnostic systems to allow for diagnostic conclusions, compared to the rather abstract assumptions of the normative model? This rhetorical question may sound *prima facie* plausible. One may reason along the following lines: the more details in a model that need to be assessed, the harder it is for the model to be fulfilled by a targeted system, so that conclusions that require a specific outcome in the assessment of a diagnostic model that is more detailed are harder to come by. If they are harder to come by, meeting these more demanding conditions should be assumed to provide better justification. However, on closer investigation this reasoning is wrong. What such reasoning actually tracks is not the internal justification of conclusions but their *informational value*, which as I will argue is not a source of intrinsic justification, since diagnostic modelling rests explicitly on indication – that is, on reliability. But before I argue along these lines, let us make clearer what I mean by informational value.

By the informational value of diagnostic conclusions, I mean the number of insights we have into a system based on a diagnosis given to it. Hence the informational value of a diagnosis equals the number of constitutive factors assumed in a diagnostic model that need to be matched with the modelled system to support a diagnostic conclusion about it. Let me give an example. Consider a certain portion of a normative model that assumes a normally functioning system to operate, so that it provides a certain output given a certain input, considers an error to be present based solely on inputs and outputs to the system. Hence the informational value of a diagnosis based on a normative model will have the informational value of precisely this aspect of the system's behaviour: the erroneous input/output. Now compare this to a diagnosis based on an error model. An error model in the context of a differential diagnostic process consists of several propositions regarding constitutive facts required to be true about the system (beyond providing a certain output given some input), that must be present in the system for the model's successful application and therefore the justification of a diagnosis based on the model. The enabled error diagnosis in this case will be more informative than a diagnosis based on a normative model, since the error model goes beyond the normative model and provides further details to assess when making a diagnostic ascription to the system. So much for informational value and why it is higher in some approaches within model-based diagnostics than in others. Now we can return to the question of whether informational value provides internal justification, and thus whether some diagnostic conclusions have better intrinsic justification than others – in our example. Let me first explain why I believe informational value does not contribute to intrinsic justification.

The informational value of a diagnosis differs based on the modelling procedure enabling it, but the differences in informational value do not translate into differences of intrinsic justification. The intrinsic justification for thinking that a given diagnosis indicates a specific error rests on the assumption that the model used for the diagnosis allows us to reliably predict the presence of the error, hence the vehicle of intrinsic justification is indication, which is constituted by a reliability relationship. This reliability, however, does not depend on the informational value of a diagnosis, hence it is justified by the reliability and not the informativeness of the diagnosis. To illustrate this, let me return one last time to the example of the barking dog and the doorbell that I used earlier when explaining indication.

This time, imagine that we have two scenarios. In the first scenario, the dog barks whenever the doorbell is rung and not when it is not rung, but it makes no difference how often or how fast the bell is rung. In the second scenario, the dog barks only if the doorbell is rung twice and not when the doorbell is not rung or is rung more or fewer times. In both cases, the dog's barking reliably correlates with a state of affairs; therefore in both cases it indicates this state of affairs. However, the two states of affairs correlating with the barking of the dog differ specificity. In the first instance, the dog barks whenever the bell rings, and in the second, there is a specific pattern of ringing whenever the dog barks. It appears that if in both scenarios we cannot hear the ringing of the bell but only the barking, and if we are familiar with the barking behaviour of our dog, we would have more detailed knowledge about the obtaining state of affairs in the second scenario than in the first. In the second scenario, the barking of the dog indicates not only that it has rung, but more specifically that it has rung exactly twice, while the barking in the first scenario may indicate any number of rings. The barking of the dog in the second scenario therefore seems to be an indicator with more informational value. We know the bell rang, and we know it did so in a specific way: exactly twice. In the first scenario, we only know that the bell rung. However, assuming that the dog's barking is indeed reliably correlated with the relevant doorbell-ringing scenarios, it appears that our reasons to believe what both barks indicate are equally well justified in both scenarios, since both instances of barking indicate what they indicate with the same reliability. Because reliability understood as correlation is the determining factor of indication, this is all that counts in this context, no matter how informative the state of affairs may be that is indicated.

If we bring this back to modelling, we may think of modelling in analogy to these scenarios. The first scenario, in which the dog's barking provides only the information that the bell has rung, may be compared with a diagnosis based on the application of a normative model, providing a rather thin understanding of the presence of an error, only in terms of input and output patterns. The second barking scenario may be thought of in terms of error models used in differential diagnostics, since it indicates not only that the bell rang, but moreover that it was rung twice. By analogy, the use of an error model to provide a diagnosis goes beyond the assessment of abnormal input and output patterns and takes into account more specific aspects of the erroneous occurrence. Just as the reliable correlational relationship between barking and the state of affairs it indicates is what allows the instances of barking to justify the belief in the state of affairs indicated by them, it is the reliable relationship between the applicability of a model and the state of affairs (the error) it indicates that provides the resulting diagnosis with intrinsic justification. Just as the barking of the one dog is not better justified than the barking of the other because of informativeness but solely because of reliability, likewise a diagnosis based on one approach to diagnostic modelling is not better justified than one provided by another because of the informativeness about a state of affairs in the model based on the respective dogs barking. Given the nature of the indication relationship consisting in reliable correlation and given the fact that, as laid out in earlier sections, diagnostic modelling is supposed to use models as indicators for the presence of errors, the intrinsic justification of diagnostic conclusions is based on the vehicle of indication and thus on reliability as the central epistemic value.

While reliability is crucial for internal justification of diagnostic conclusions, it is not the whole story regarding justification. For internal justification, as presented here, to bear any general epistemic weight, we have to be convinced that a diagnostic system in model-based diagnostics that is able to justify things internally is also justified on a more fundamental level. We must be convinced that it not only provides us with an epistemically plausible way to think about conclusions as being justified within the system, but that the framework itself based on which these inferences are made is valid. The diagnostic system requires external justification. We may well have a diagnostic system being used in model-based diagnostics that provides us with reliable (i.e., repeatable/stable) results, but we also need a reason to be sure that these results indeed track the presence of actual errors, a reason to believe that the outcomes are valid – a reason to believe that they really identify the presence of specific errors.[32] Showing how a diagnostic system (a set of diagnostic models used to address a certain system) in model-based diagnostics gains external justification is my next step.

2.4.4.2 External Justification

As discussed in detail in section 2.3, the basic approach of diagnostic modelling is to apply diagnostic models (normative models or error models) to a diagnosed system and use the result of the comparison procedure. The results are the identifiers of matches and mismatches between these models and the real-world system, and they are used as indicators for the presence of suspected errors. Therefore, as discussed in the previous subsection, the internal justification of diagnostic conclusions in diagnostic modelling rests on the assumption of the reliability with which the use of these models allows us to indicate the presence of the targeted errors. However, as I mentioned repeatedly, to assess whether the modelling used indeed reliably indicates a targeted phenomenon (i.e., a specific error), we need an additional source of justification. We need some external justification that provides us with reason to believe in the diagnostic results by ensuring the validity of the models that are used for purpose of indication. This would mean, for example, justifying that an error model indeed contains relevant constitutive factors of an error. Only then does the model seem a legitimate basis for an inference to best explanation regarding the presence of this error in a differential diagnostics procedure, hence making it permissible to use its applicability as an indicator for the presence of this error. If the need for this additional dimension of justification is acknowledged, the questions becomes: how do we gain this external justification for the validity of diagnostic models so that they can be assumed to be valid tools for use in the inferential machinery of diagnostic modelling producing externally justified diagnostic conclusions?

I argue that the external justification of diagnostic conclusion *qua* use of valid models in the diagnostic process depends on the justificatory strength of the background theories used to set up diagnostic models. To show why, we must compare the standard approach of modelling as presented in section 2.1 and the more specific use of modelling for diagnostic purposes. As discussed in section 2.1, in the attempt to develop a model of a system that allows for matching and simulation of

32 The relationship between reliability and validity that I presuppose here is the commonsensical understanding of the relationship between reliability and validity in measurement. Reliability of a measurement depends on its validity, whereas validity does not depend on reliability, and a valid measure is generally reliable (Bajpai and Bajpai, 2014).

certain features of a system, a modeller can use a range of sources to inspire their model structure. Modellers may be inspired by their intuition, draw on other models whose structure they deem promising to target the intended system's features, or capitalise on pre-existing theories of the phenomenon being modelled. A model structure derived from these sources will then usually be tested against the targeted real-world system in model/world comparison to see whether the model matches the real-world system well enough in terms of representational and dynamic fidelity. If not, the model will be revised until it does. Having a model successfully go through this process seem to justify the assumption that the model is accurate enough for the purposes of the modeller to be accepted as a model of the targeted system. If we look at the process of diagnostic modelling, however, there seem to be some differences.

Diagnostic modelling differs in some regards from the standard procedure for modelling a system that I have just sketched out. It does by virtue of its epistemic purpose. While modelling as just described aims to provide a good representation of the targeted system, diagnostic modelling attempts to make a judgement about the targeted system. The former kind of modelling takes the real-world system as a benchmark for the model and thus derives its legitimacy as a model by displaying model/world comparison. Diagnostic modelling, by contrast, takes its models as benchmarks to test the real-world system regarding features that suggest a diagnostic conclusion. If the diagnostic modelling process itself cannot equip the models used within it with plausibility, but requires their legitimacy before they are applied, then there are arguably ways to ground trust in these models.

The ways for diagnostic models to claim validity comes down to two options. The first option is that in the step of model construal, the diagnostic model is set up based on a background theory that provides a constitutive explanation either of what the normative behaviour of the system should look like (normative model) or of the constitutive factors of specific instances of error in a system (error model). The justification of the assumption that these models indeed capture the relevant constitutive factors of the system if a diagnosed condition is present then itself depends on the quality of the theories from which these models are derived. However, the question of when the acceptance of a theory is justified is in itself a highly controversial question that I will not be able to explore; what I can do here is merely to clarify that this is where the burden of justification shifts to. The second option is that the diagnostic models used are themselves results of earlier modelling attempts that were not diagnostic modelling, but rather system modelling, either focusing on normal system behaviour and its constituents in order to develop models for it or else aiming to provide such models for errors or abnormally behaving systems. Once these models have been developed, they can be reused by diagnostic modellers for their own purposes. As we will see in the next chapter, in psychiatry most diagnostic models rest either on our folk-psychological theories of human psychology and behaviour or on psychopathological research. For now, however, let us consider what is implied by these two ways for diagnostic model to gain plausibility.

If diagnostic models are derived either from theories providing the relevant constitutive explanations or from pre-established constitutive models, diagnostic conclusions enabled by the use of these diagnostic models can enjoy external justification. Diagnostic models that are used in model-based diagnostic procedures can enjoy support that grants them plausibility, so that their application to identify erroneous conditions seem to be justifiable. I say "seems", because these models will of course be only as plausible as the theory they are derived from or the quality of the modelling process that produced them. However, if these sources suffice, which is an empirical matter that would have to be evaluated for any given diagnostic system, it seems that diagnostic conclusions arrived at by the diagnostic modelling enjoy external justification. They enjoy external justification partly because the conclusions drawn within the system are a *supposedly* reliable way to become aware of errors in the system. But indeed, we have reason to believe that these inferences are also reliable in that they are based on adequate diagnostic models that are able to identify sufficiently relevant constituents of a targeted error, which in the presence of these constituents *qua* inferences to the best explanation allow us to take the applicability of these models as an indicator of the presence of the relevant error. External justification – that is, the validity of the models to be used as reliable indicators of error – would thus rest on the quality of the background theory and modelling approaches used to come up with the diagnostic models in the first place. Thus, the principal source of external justification for diagnostic conclusions is now laid out.[33]

---

33 This idea that the validity of diagnostic models depends on their capacity to pick out the constitutively relevant aspects of systems, enabling them to actually pin an error label to an underlying constitutively responsible makeup in the diagnosed system, is related to the understanding of validity in of test instruments. Test instruments are usually judged to be valid when they actually measure the construct at stake – a use of the notion going back at least to Kelley (1927). When is this "actually measuring" requirement satisfied? There are causal as well as correlational proposals. Borsboom and other psychometricians (Borsboom, Mellenbergh, and Van Heerden, 2004; Borsboom, 2005) proposed that a measuring procedure "is valid for measuring an attribute if and only if (a) the attribute exists and (b) variations in the attribute causally produce variations in the outcomes of the measurement procedure" (Boorsboom, Mellenbergh, and Van Heerden, p. 1061). Many philosophers (e.g., Angner, 2011; Cartwright and Bradburn, 2001; Alexandrova, 2017; Michel, 2019), however, substitute b) for a mere correlational criterion. My understanding of validity would fit well with a correlational proposal. Note, however, that I do not mean to claim that model-based diagnostics is a measurement process; to evaluate whether this is true or not would require work that is beyond the scope of my project. All I was interested in here is giving a better grasp on the idea of validity.

2.5 Conclusion

In this chapter I have presented my account of models and modelling, which in its application to psychiatric diagnostics will provide my answer to the Methodological Question: What is the method of psychiatric diagnostics? Namely, that psychiatric diagnostic reasoning is qualitative, constitutive diagnostic modelling. The content of this chapter enables us to understand what qualitative, constitutive diagnostic modelling is and will provide the basis on which to formulate what needs to be shown about diagnostic reasoning to make plausible that it embodies this kind of modelling. Doing so will be the task of the next chapter. To end this chapter, let us briefly review what has been done.

In this chapter, I first introduced a general understanding of modelling as a process, understood as what is called the indirect strategy of representation. Next, I presented some specifications of modelling: one specification regarding a potential format of modelling, qualitative modelling, and one specification regarding a goal-driven epistemic approach one may take when using modelling – namely, diagnostic modelling. Thinking in terms of the Methodological Question, these parts of the chapter provided the description of the method I claim to be used in psychiatric diagnostics. Next, I discussed the inferential strategy pursued by diagnostic modelling, the constitutive indicator strategy, followed by an exploration of the inferential patterns that underlie the inferences generated via this route: inferences to the best explanation and apophatic inferences. Finally, I discussed the justification of conclusions drawn in model-based diagnostics. Again, thinking of the Methodological Question, this second part of the chapter provided material with which to address its remaining aspects, beyond the task of *describing* the method in place in diagnostic reasoning. The second half of this chapter provided the rationale behind the procedures of model-based diagnostics as a method as well as an understanding of how its conclusions are supposed to be deemed justified.

With all three aspects of an answer to the Methodological Question (description, rationale, and justification) in hand at the end of this chapter, the remaining task is to show that the method of diagnostic reasoning is indeed at work in diagnostic psychiatric reasoning and thus that the methodology presented here applies to it. This brings us to the next chapter, in which this task will be completed.

3. Diagnostic Reasoning as Modelling

In the previous chapters, I offered an overview of the core practices of clinical psychiatric diagnostics (Chapter 1) and presented a methodology for qualitative, constitutive diagnostic modelling (Chapter 2). The separate presentation of these two topics has paved the way for my next step in this chapter. Here, I argue that the process of diagnostic reasoning that psychiatrists engage in during clinical psychiatric diagnostics can be understood as a qualitative, constitutive diagnostic modelling procedure plus an additional layer of processing that should be understood as pattern recognition. Thus I establish my model-based account of psychiatric diagnostic reasoning.

To make plausible that psychiatric diagnostic reasoning can be understood along the lines of my proposal, I will show that the central features of the method described in Chapter 2 maps the diagnostic procedures described in Chapter 1. I will demonstrate that the inferential processes spelled out by the method of diagnostic modelling make for a plausible proposal of how to understand the steps of diagnostic information-processing.[1]

Considering my in-depth discussion of modelling in the last chapter, I propose that the following aspects of qualitative, constitutive diagnostic modelling must be shown to be present in psychiatric diagnostics to make the method plausibly present in psychiatric diagnostics, as well as descriptively adequate and suitable for provide an understanding of the inferential processes of diagnostic information-processing.

The first three criteria derive from the general understanding of modelling (2.1), which follows the three-step procedure of model construal, analysis, and model selection.

1 As discussed in 1.3, diagnostic information-processing as part of the diagnostic process was not included in the descriptive proposal of Chapter 1 since there is no widely upheld understanding of these aspects of the diagnostic process. Rather, an understanding of psychiatrists' processing of diagnostic information is part of what any answer to the Methodological Question must provide a plausible proposal for – one that makes sense of and is constrained by the inputs and outputs to these instances of information-processing.

i) Construe theoretical structures intended to represent the target based on little previous knowledge about the actual target system (construal)
ii) Consider the regularities of the model structure(s) that have been set up (analysis)
iii) Engage in a fidelity criteria-based selection process in which the model(s) are compared to the real-world system and a choice is made to accept or reject the model (model/world comparison and in result model selection)
iv) The model structure(s) must consist of elements and relationships specified in qualitative terms, more specifically as linguistic propositions.

If psychiatric diagnostics can be understood as any kind of modelling process, there should be steps in the diagnostic process that are plausibly interpretable as these three steps of the modelling process. In addition to these criteria, which make it plausible to think of psychiatric diagnostics as modelling at all, I add a fourth criterion to be fulfilled, which makes plausible that diagnostic psychiatric reasoning is, more specifically, qualitative modelling.

v) The model structure(s) must consist of elements and relationships specified in qualitative terms, more specifically as linguistic propositions.

Finally, to establish that psychiatric diagnostics is diagnostic modelling and constitutive modelling as described in the last chapter, the following criteria should be met.

vi) The modelling procedure employs a normative model of the diagnosed system to indicate which outputs of the system qualify as abnormal and classify them accordingly as being at least prima facie errors
vii) The model structure(s) used as error models within the diagnostic process are constitutive models

The diagnostic modelling process may employ either simple normative-model diagnostics or error-model based differential or exclusion diagnostics.

If all these criteria can be shown to be met in psychiatric diagnostics, it seems that the mapping between qualitative, constitutive diagnostic modelling and psychiatric diagnostics holds. This would entail that in considering this method, the further methodological considerations about diagnostic reasoning that I supplied in Chapter 2 can be applied to psychiatric diagnostics. Demonstrating this, plus adding some remarks on the pattern-recognition based stage of psychiatric diagnostics that I will claim to apply in the step of disorder diagnosis selection, will establish my answer to the Methodological Question: the model-based account of psychiatric diagnostic reasoning. My answer will claiming that psychiatric diagnostics is largely diagnostic modelling process plus one level of pattern recognition. To discuss how clinical diagnostics meets these conditions related to my proposed method of modelling, and what role pattern recognition plays on top of it, I will proceed as follows.

In the first section (3.1), I will look at the initial screening phase of the diagnostic process and its preparation of the in-depth evaluation, and show what criteria of modelling are fulfilled in it. For this, I will introduce three clinical examples and go into more clinical depth than in the first chapter of this book, which was intended primarily to provide a conceptual grasp of each step of the process of psychiatric diagnostics. My examples will be the complaint of reluctant speech, the complaint of constant worrying, and the complaint of relationship problems. I will discuss what potential in-depth diagnostic evaluations these complaints would entail, and what psychiatrists would be interested in when evaluating whether these complaints constitute psychopathological symptoms. Following the presentation of these examples, I will argue that the screening step of the diagnostic process equals a normative-model based *prima facie* error recognition (corresponding to criterion v above) and that the *prima facie* error recognition leads to the diagnostic information-processing that prepares the in-depth evaluation. This evaluation seems to consist of construal (*i*) and analysis (*ii*) of models that are qualitative (*iv*), and constitutive (*vi*) in nature and that serve as diagnostic models. In this way, I will have shown that the diagnostic process meets several of the above-mentioned criteria for it being a qualitative, constitutive diagnostic modelling process (specifically, criteria (*i*), (*ii*), (*iv*), (*v*), and (*vi*)).

Next (3.2), I will take some of the clinical examples discussed so far and ask how to understand the execution of the diagnostic in-depth evaluation that was prepared for in the screening phase. I will argue that carrying out this in-depth evaluation means performing a model/world comparison (*iii*), and will show how this instance of model/world comparison is realised as a form of diagnostic information-processing that equals the diagnostic modelling process of differential or exclusion diagnostics (*vii*). This section will show that the two remaining criteria are met (*iii*, *vii*) and thus that the aspects of psychiatric diagnostics discussed so far can rightfully considered to be qualitative, constitutive diagnostic modelling.

Once this mapping between the diagnostic process and diagnostic modelling is established, I will come to the part of diagnostics not exhausted by modelling. In a third step (1.3), I will use my case examples to discuss how the present psychiatric symptoms identified in the in-depth evaluation of the patient (or as a result of the diagnostic model/world comparison process) are then used to set up the case formulation. The case formulation is a representation of the modelling outcomes, summarising the results explaining its outcomes and thus informing the syndromal disorder diagnosis that is made in accordance with a chosen diagnostic manual. This step of diagnostic information-processing, I will argue, is a pattern recognition process performed by experienced clinicians.

Finally (1.4), I will conclude this chapter by summing up how the proposal thus presented answers the Methodological Question. In Chapter 4 I will then move on to discuss additional desiderata for an answer to the Methodological Question and how my answer is also fulfilling those.

3.1 Screening as Modelling

In the opening phase of clinical information-seeking stands the question: what are the reasons for the patient's admission to the psychiatric institution? This question is initially answered as a result of the part of the diagnostic information-gathering process that functions as screening for complaints. This information will usually be attained directly from the patient during the first clinical encounter by asking for the reasons why he requires psychiatrist services (in the initial phase of the psychiatric interview) and by making initial observations of the patient's behaviour and asking more specific questions and potentially tasks (via mental status examination and testing). Admission charts and family reports may also be used for this purpose. This minimal initial information about the patient provides the psychiatrist with her initial screening impression of the patient and his complaints, providing her with a list of complaints that may qualify as psychiatrically relevant complaints. Such a list might, for example, contain information about the patient reporting "sleep problems", "feeling sad all the time", having "lost pleasure in free time activities", "feeling tired all the time", "worrying a lot", as well as behaviours that give the impression of being potentially psychopathologically relevant – for example, that the patient appears to have "problems concentrating on his actions and the conversation", "shows increased psychomotor activity by rubbing and kneading his hands and chewing his nails", or is "remarkably reluctant and laconic in speech". Thus, complaints may entail subjective reports as well as second- or third-person observations. From such list of complaints, the psychiatrist generates ideas that might explain the patient's complaints. If she sees an option for how one of these complaints might constitute a psychopathological condition, she will further explore the patient's condition to decide whether or not this is the case. Let me make this beginning of the diagnostic process, which I described in its general format in the first chapter, more concrete by discussing the diagnostic procedure for two of the three aforementioned complaints that I will use throughout this chapter to illustrate my argument: the complaints of reluctant speech and constant worrying. I will discuss what initial considerations a psychiatrist may use for their potential in-depth evaluations, as well as what evaluations a psychiatrist may look to carry out in order to determine whether the complaint is a psychopathological symptom, another medical problem, or a distressed but non-pathological state of mind.

3.1.1 Examples of Screening

For the first example, consider the fact that the psychiatrist recognises a patient's unusual speech pattern in the course of their conversation. In the table, a noticeable speech pattern in the left column is contrasted with the normally expected pattern in the right column.

Table 3: Noticeable speech pattern (left). Normal speech pattern (right)

Noticeable	Normal
Psychiatrist: Good Morning, Mr Jones. What can I do for you?	**Psychiatrist:** Good Morning, Mr Jones. What can I do for you?
Patient: Help.	**Patient:** I came to you because I have some problems that I think I need help with.
Psychiatrist: And I will try my best to do so. Can you tell me something about the reason you are reaching out for help?	**Psychiatrist:** And I will try my best to do so. Can you tell me something about the reason you are reaching out for help?
Patient: Yes.	**Patient:** Well, thanks. I feel sad and empty, and I do not know what I should do about it. It started [...].

If an interview conversation goes on like this, and the patient's language production remains so remarkably laconic, the psychiatrist will come up with the idea that the patient may suffer from a psychiatric condition called alogia. According to Sadock and Sadock. (2008, p. 27), alogia is a "laconic speech condition characterized by a reduction in the quantity of spontaneous speech; replies to questions are brief and unelaborated, and little or no unprompted additional information is provided. Occurs in major depression, schizophrenia, dementia, or schizotypal personality disorders (APA, 2013, p. 817). Alogia is also called "poverty of speech".

As a second example, let me come back to the potential self-description of a patient saying "I worry all the time." If a patient reporting such an indistinct complaint indeed turns out to indicate a psychiatric symptom, there is more than one option for which one it might be. It could be *generalised anxiety*, which is usually understood as "[c]hronic, excessive and uncontrollable worry about multiple topics" (Hirsch et al., 2013, p. 388), or a more specific object/situation-related anxiety of psychopathological value, which would be of similar character but tied to a frequently occurring trigger. Alternatively, this "worry" might also turn out to be a form of compulsive thought that is causing negative emotions in its evaluation. In a clinical context, this could be understood as a specific kind of unwanted, unintended, recurring, and intrusive cognitive event whose content is experienced as egodystonic but subjective – that is, a product of the individual's own mind.

Considering the patient's complaints, the psychiatrist will draw on her background knowledge to consider alternative explanations for those that are initially recognised as plausibly indicating psychopathologically relevant problems (such as the two I am offering as examples). Through further evaluation of the patient, she will decide whether the *prima facie* psychiatrically relevant complaint represents a psychiatric symptom, a non-psychiatric medical symptom, or maybe even no medically relevant symptom at all.[2] What may be the various alternative options that the psychiatrist has in mind in her examination of the initial complaint that might speak for one (or another) psychiatric symptom or the alternatives?

Let us again begin by considering the potential case of alogia. Given the observed complaint, several diagnostic hypotheses may come to mind. Each maps onto a diagnostic outcome; some speak for the patient's behaviour being the psychiatric symptom of alogia, while others may suggest alternative medical diagnostic conclusions, or that the complaint has no symptom value at all. The psychiatrist might theorise that:

2 At this point, the question may arise as to whether evaluations of initial complaints of patients in a psychiatric context always include the option of turning out to be only a *prima facie* psychiatric complaint – i.e., to be a non-psychiatric medical problem or not a medical problem at all. One reason to doubt this has been presented to me by a colleague is a patient reporting hearing voices most of the time for some weeks. How could this not be a psychiatric problem? Before I respond to this problem, let me provide a more general answer. It might be possible that there are initial complaints that allow only for an assessment that shows them to be psychiatric. In this case, a further evaluation beyond the recognition of the complaint would not be necessary. Such cases, which would be an exception that I would have to tolerate, are *possible*, but whether they *exist* is another question. I am not aware of such cases, and so I consider them to be at least rare. Now let me come back to the hearing-voices case. It might be that this patient hears voices because they suffer from the psychiatric symptom of hearing voices, which is primarily associated with disorders on the psychotic spectrum but can also occur in depression, for example. However, hearing voices can also result from lesions of acute inflammation of the brain (Silva and Brucki, 2010) and can occur during sleep deprivation and starvation. Even for the initial complaint of hearing voices, therefore, there are explanatory options to evaluate it that would lead to the diagnostic conclusion that the symptom is a non-psychiatric medical complaint or not a medical complaint at all. To make this clear, think of a psychiatrist who is checking the necessary criteria for providing a diagnosis of schizophrenia. Whether or not this diagnosis can be provided depends on the question of whether the patient hears voices. If the patient hears indeed voices but the psychiatrist has good reason to suspect that this is due to lack of sleep, she apparently should not and will not make the diagnosis, since psychiatric diagnosis usually includes those alternative explanations for diagnostically relevant features.

A. The patient did not want to consult the psychiatrist but does so to satisfy relatives or friends who pressure him to do so.
B. The patient might have an unusually pedantic way of speaking, not associated with any morbid condition.
C. The patient may have taken drugs impairing his language-related cognition – e.g., cannabis (Dellazizzo et al., 2022).
D. The patient might have had a traumatic brain injury (TBI) that could have led to this condition.
E. The patient might suffer from specific cognitive deficits in language processing responsible for his speaking behaviour.

Considering this list of possible explanations, they can be matched with the diagnostic outcome that their truth would support. If hypothesis A. were true, the patient's language production would not be a sign of psychiatrist or medical problems; it would be a motivated behaviour expressing his lack of interest in cooperating with the psychiatrist. If B. applied, this would again not be the psychiatric symptom of alogia but rather someone with an unusually pedantic way of speaking – something that happens from time to time and may lead to misdiagnosis. This is a problem that, as the literature indicates (Andreasen, 2016), has been observed particularly in interaction with administrators, politicians, scientists, and (perhaps unsurprisingly) philosophers. If C. applied, the patient's laconic speaking behaviour would be considered a medical problem, namely a temporary drug-induced cognitive alteration of language behaviour, which again is not a symptom of a psychiatric disorder, only an effect of a momentary intoxication.[3] If D. applied, the patient's problem would be considered a medical problem falling under the specialisation of neurology, but

3 At this point you may wonder, why not consider a substance-induced mental alteration (e.g., under the influence of cannabis) that causes an acute speech impairment alogia. Or, as one may ask more generally, why should two hypothetically similar token behaviours or mental states be classified as a psychiatric symptom or sign on one occasion, but as non-psychiatric on another? This is due not to some strong metaphysical distinction, but rather to the special place that psychiatric symptoms and signs currently have in medical semiology (Altable, 2012). In medicine, symptoms are traditionally considered manifestations of a disease, or, to put it in more philosophical terms, they are representations of the presence of disease, and therefore of physiological alterations considered causally responsible for their presence. If a symptom or sign is caused by a disease condition that is not considered a mental disorder, it is, for the purpose of providing diagnoses of psychiatric disorders, not considered to be a psychiatric sign or symptom. This does not mean that, in the end, research might not show that part of the causal pathways responsible for the occurrence of the symptoms is shared by a psychiatric disorder and a disease with similar psychological or behavioural symptoms. In consequence, if a psychiatrist is convinced that the alogia-like change of language production is best explained as a result of the patient's recent consumption of a substance, they will mention the patient's state but not consider this impairment for the further psychopatho-

would again not mean that the patient's complaint would be considered a psychiatric symptom. Based on our current best understanding of alogia as a psychiatric symptom, which I will discuss in detail later, only if hypothesis E. provides the best explanation for the patient's language behaviour will the patient be considered to suffer from alogia as a psychiatric symptom.

In the same way as for the potential case of alogia, the psychiatrist would come up with a list of options to address the complaint about constant worrying:

a. The patient's worry may be the result of an increase of arousal occurring in response to dealing with current high-stress or hostile circumstances.
b. The patient takes medication or drugs on a regular basis that, depending on the dose, can cause anxiety reactions (e.g., corticosteroids or caffeine)
c. The patient's constant worrying turns out to consist in thoughts coming to his mind whose content is not particularly distressing but that cause higher-order distress because of their undesired persistence and their negative appraisal.
d. The patient's worry results from the anticipation of or reaction to a specific frequently occurring stimulus (e.g., a type of situation or object) that he is afraid of to a degree that seems extraordinarily high given its nature.
e. The patient's worry is a specific stimulus-independent reaction to expectation of unlikely menacing events and more likely but unthreatening events.

Again, the different ideas as to how to explain the complaint of the patient would, if they applied, lead to different diagnostic judgements. If a. applied, the patient's

---

logical evaluation that feeds into the ascription of a mental disorder – that is, they will not consider it as a symptom of a mental disorder.

Take an example: if the decision as to whether the reluctant speech of a person who has consumed cannabis is considered alogia or not is the tiebreaker in whether the psychiatrist will diagnose schizophrenia, and the psychiatrist has good evidence that the patient does not show this impairment if they are not intoxicated, the psychiatrist would not diagnose schizophrenia. Why not? Because the patient's condition is by definition not a sign of psychiatric disorder; it is substance-induced, and as such, it has a distinct aetiology that in itself does not directly entail a mental disorder (it is a potential addition that would play no role in the diagnostics here). For this reason, many diagnostic manuals offer specific categories for instances of impairment or alteration of cognition and experience specifically as substance-induced. Note that this is not to say that substances may not in the end cause conditions that in themselves will qualify as psychiatric, neurological, or other medical disorders. For example, long-term consumption of alcohol may cause the development of Korsakoff syndrome, which is considered an irreversible form of Wernicke encephalopathy leading in particular to impaired retrograde and anterograde memory and confabulation (Covell and Siddiqui, 2022). The consumption of a range of substances may contribute to the onset and substance-independent persistence of psychosis (Deng et al., 2012).

complaint would be a normal psychological response to current and ongoing life circumstances. If b. applied, the patient's complaints would again be considered not a psychiatric symptom but a side-effect of medication or other substance-induced complaint. If c. applied, the diagnosis might be that instead of suffering from a psychiatric anxiety symptom, the patient suffers from persistent compulsive thoughts, another psychiatric symptom that causes distress. If d. applied, the patient's complaint would be considered stimulus-specific psychopathological anxiety, and finally if e. applied, the complaint would turn out to be a general psychopathological anxiety reaction. So much for the available options for diagnostic evaluations of the initial complaints. However, a well-trained psychiatrist is not only able to come up with the two lists of hypotheses addressing the complaints; she also possesses a knowledge base regarding how to evaluate each hypothesis. This brings me to the next topic of this subsection, namely the considerations undertaken by the psychiatrist as to how to evaluate the diagnostic options in the next step of the diagnostic process: the in-depth evaluation.

To know what to do in the in-depth evaluation, the psychiatrist calls upon their knowledge about what diagnostic information would have to apply to the patient with the complaints in question, to support each of their optional evaluations. We can think of the assumptions that should be true in the case of the patient as a set of interrelated propositions that the psychiatrist can evaluate against to generate diagnostic data about the patient's presentations. These sets of propositions result from the background knowledge in psychiatry (including the predisposing, aetiological, maintaining, and co-occurring factors for psychopathology), general medical background knowledge, as well as folk-psychological understanding of human minds and behaviours. To illustrate what these set of propositions may look like, let me come back to the two complaints and their potential evaluations and expand on three potential diagnostic evaluations of each complaint and what the psychiatrist might look for to verify them. Again I shall begin with the case of reluctant speech.

To discuss the setup for the in-depth evaluation of the potential case of reluctant speech, let us consider three of the aforementioned diagnostic options and what a psychiatrist might look for to validate them. Let us take the non-pathological case of pedantic language use (B.), the actual case of disturbances of language control retrieval in which the patient's complaint would be evaluated as psychiatric symptom alogia (E.), and the case in which the patient's language problem would lead to the evaluation of being a non-psychiatric but medical problem deriving from the option that the patient suffered from traumatic brain injury (D.). Let me start with option B.

To evaluate whether the patient showing the complaint of reluctant speech may just have an atypically reluctant manner of speaking that is normal for the patient (B.), the psychiatrist may set up the following set of propositions:

- The patient can elaborate their answers if asked to.
- The patient recognises that their answers are unusually short and can justify their manner of speaking by explaining their motivation (e.g., wanting to save the doctor's time, or wanting to be as precise as possible).
- The patient can report that his way of presenting information is not something that has developed recently but is rather their normal way of conveying information.

If possible, the psychiatrist also speaks to relatives, friends, or other medical professionals to verify the statement that:

- People who know the patient report that the patient has always tended to speak this way.

If, on the other hand, the psychiatrist wished to evaluate the hypothesis that the patient's language behaviour resulted from a traumatic brain injury (TBI)[4] (D.) they would evaluate the following propositions:

- The patient recently took some sort of blow to the head (e.g., by falling or having an accident).
- The patient did suffer some such blow, and lost consciousness or had loss of memory of events immediately before or after the blow.
- There were alterations of mental states at the time of the accident (e.g., feeling dazed, disoriented, or confused).
- Lesions that indicate traumatic brain damage be seen in computed tomography.

Finally, let me come to the case in which the psychiatrist would like to assess whether the patient's complaint presents a specific language-processing disturbance (E.) that would render their complaints a case of the psychiatric symptom of alogia. For this, let me say a bit more about our current best understanding of alogia in psychiatric sciences.

The chief cognitive impairment behind alogia in psychiatric cases involves an impairment of *control retrieval* – part of the executive functioning that enables the retrieval of information from memory. Alogia occurs when the information is not automatically retrieved, or when there is more than one potential piece of information matching the search profile (Wagner et al., 2001; Doughty and Done, 2009; Docherty, Berenbaum, and Kerns, 2011). If a test of speech production, conducted on a coop-

4 Information about TBI and its evaluation can be found in National Academies of Sciences, Engineering, and Medicine (2019).

erative patient, shows patterns indicating this kind of cognitive impairment, diagnosis of alogia seems warranted. How to evaluate whether such a condition holds?

As already mentioned, alogia is considered a condition deriving from a disorder of the cognitive function of *control retrieval* – a part of the executive function that enables the retrieval of information from memory when either that information is not automatically retrieved or when there is more than one potential piece of information matching the search profile. This cognitive function can be tested with verbal fluency tasks. Such tasks require subjects to follow a production rule in voicing words. They may be required, for example, to say words beginning with a certain letter (testing word–letter fluency) or falling into a category such as *animals* (testing word fluency). More specifically, when being tested for alogia, an individual would be asked to produce lexical items for a certain span of time. If an individual suffers from a cognitive impairment of control retrieval, there is an increased mean response latency between each reported word when asked to produce words in a category. If the individual does not show this deficit, this suggests the impairment of other language-related cognitive functions that, in principle, could also lead to the clinical presentation. These other impairments include disorganised semantic memory (which would lead to poorer performance in category fluency relative to letter fluency) or context processing (which should lead to a decrease in the proportion of correctly reported semantically-related words) (Docherty, Berenbaum, and Kerns, 2011). If verbal fluency testing of the patient meets this prediction, it may be reasonably concluded that the patient's complaints are an instance of the symptom of alogia.

In accordance with these insights into the underlying psychology of alogia, the psychiatrist may put forward a proposition that can be evaluated during an in-depth evaluation in cognitive testing, as described above:

- The patient shows relevantly worsened outcomes in a verbal fluency task.
- The patient shows no impairment in semantic memory.
- The patient shows no impairment in context processing.

Now to the second example I wanted to discuss: the complaint of constant worrying. For this, let us again consider three of the aforementioned evaluations that a psychiatrist may have in mind: constant worrying in the context of permanent arousal due to constantly present stressful circumstances, which would not suggest a psychiatric or non-psychiatric medical problem (a.); constant worrying consisting of thoughts that are not particularly distressing in their content but that cause higher-order distress because of their undesired persistence and their negative appraisal, which would suggest the evaluation of these thoughts as persistent compulsive thoughts, rather than as an anxiety symptom (c.); worrying as a specific stimulus-independent reaction to an expectation of unlikely menacing events and more likely but unthreatening events, which would suggest that the patient's complaint indicates the presence of a psychopathological general anxiety reaction (e.).

To evaluate whether option a. applies, meaning that the patient does not suffer from a psychiatric or any other medical problem regarding the complaint or constant worry, the psychiatrist would have to evaluate whether:

- The patient is currently in a highly stressful or hostile life situation (e.g., currently being threatened and followed by an ex-partner, or having lost his job and being in significant debt and about to become homeless) that makes the worry reaction seem appropriate.
- The patient's increase in worry coincides with the occurrence and duration of the stressful life circumstances.
- The patient's worries directly concern the source of worry, or the topic of worry is closely linked another worry or a hostile experience. Alternatively, the worry may concern an occurrence that while under normal circumstances would be no problem, is experienced as being an issue because it comes "on top" of the actual severe problems that cause other, primary worries.

Next, let us turn to the case of evaluating whether option c. applies – that is, whether the patient's constant worry is a case of compulsive thought. Compulsive thoughts are an instance of the larger class of psychiatric symptoms that are called intrusive thoughts. Intrusive thought is "any distinct, identifiable cognitive event that is unwanted, unintended, and recurrent. It interrupts the flow of thought, interferes in task performance, is associated with negative affect, and is difficult to control" (Clark and Rhyno, 2005, p. 4). This class of cognitive events contains many psychiatric symptoms, distinguished partly by their content and partly by additional formal features already pointed out by Beck and colleagues (Beck, 1967, 1987; Clark and Beck, 1999) and since then investigated by several researchers (e.g., Rachman, 1978, 1981, 1997, 1998, 2003; Dougall, Craig, and Baum, 1999; Langlois, Freeston, and Ladouceur, 2000a, 2000b; Clark and Rhyno, 2005; Morrison, 2005; Romero-Sanchiz et al., 2017). Other types of intrusive thoughts are ruminative (thoughts concerning personal loss or failure), often seen in depression; intrusive memories, often seen in PTSD; worrying (dealing with threat and vulnerability), often seen in generalised anxiety disorder; hypochondriac fear, as a specific form of an anxiety; and, arguably (Morrison, 2005), thought insertion, as experienced by psychotic patients.

To evaluate whether the patient suffers from frequently occurring compulsive thoughts whose occurrence makes the patient worry about them due to their appraisal, and which induces shame and may damage the patient's self-image, the psychiatrist must evaluate whether what the patient calls constant worry is indeed tied to the phenomenon of compulsive thought. This can be evaluated by checking whether the patient has thoughts that:

- are distinct thoughts, primarily experienced as visual (i.e., visual mental imagery) entering conscious awareness
- are attributed to an internal origin (i.e., the patient assumes ownership of these thoughts)
- are considered unacceptable or unwanted due to their egodystonic nature (i.e., their content is inconsistent with the subject's self-image or moral convictions)
- are evoking significant feelings of shame
- are interfering in ongoing cognitive and/or behavioural activity
- are unintended and nonvolitional or have wilful independence
- are recurrent or repetitive
- are difficult or impossible to control or dispel
- arise more frequently under increased stress

Finally, let me come to the third diagnostic option (e.). This interpretation would be that the patient's complaint of constant worry turns out to be the symptom of generalised anxiety, which is a psychopathological form of worry. Worry, if considered as a psychopathological symptom, can be understood as a "chain of thoughts and images, negatively affect-laden and relatively uncontrollable. The worry process represents an attempt to engage in mental problem-solving on an issue whose outcome is uncertain but contains the possibility of one or more negative outcomes" (Borkovec et al., 1983, p. 10; Sibrava and Borkovec, 2006, p. 1).[5] More particular features making pathological worry identifiable by clinicians have been discovered and replicated in a wide range of research on pathological generalised worry (e.g. Borkovec and Inz, 1990; Wells and Morrison, 1994; Wells, 1995; Clark and Claybourn, 1997; Stöber, 1998; Wells et al., 1999; Stöber et al., 2000; Langlois, Freeston, and Ladouceur, 2000a, 2000b; Hoyer et al., 2001; Stöber and Borkovec, 2002; Ruscio 2002; Lee et al., 2003; Ruscio and Borkovec, 2004; Watkins et al., 2005; Sibrava and Borkovec, 2006; Hirsch and Mathews, 2012; Hirsch et al., 2013). The understanding of pathological generalised worry emerging from this research suggests that it:

- predominantly takes the form of verbal reasoning
- is non-specific, abstract, or general in content (e.g., "what if *the worst* happens?")
- is persistent (i.e., of long duration)
- is closely linked to the individual's current concerns

5 Although at first glance perhaps similar to rumination, another psychiatric symptom, there are relevant and clear differences, most prominently regarding contents. Cognitive phenomena labelled as worry concern thoughts of possible future threats impinging on the individual. Rumination, on the other hand, is usually associated with thoughts whose contents concern past negative events or negative personal attributes (e.g., Watkins et al., 2005; Hirsch et al., 2012).

- is attributed to an internal origin (i.e., the patient assumes ownership of these thoughts)
- is experienced as egosyntonic
- is difficult or impossible to control or dispel
- entails a stress-inducing faulty appraisal concerning whether the feared consequences might come to pass ("worry about worry")
- is thought to have the positive power to potentially prevent the feared event

Accordingly, these would be the features that the clinician would look for in a patient to support the diagnostic assessment that the patient suffers from pathological generalised worry.

So far, I have discussed examples of complaints whose potential evaluation is categorical: either the patient suffers from alogia or not. Such categorical decisions about symptom attributions are significant, since in later stages of the diagnostic process, the absence or presence of this symptom will contribute to determining the symptom-based disorder diagnosis. However, in more recent editions of diagnostic manuals, such as the DSM-5 (APA, 2013) and the ICD-11 (WHO, 2019), the identification of symptoms as either present or absent has been supplemented with dimensional ratings.[6] How does this work? To answer this question, let us take the example of personality disorders, which have seen the most pronounced developments in terms of dimensional diagnostics.

In the new DSM-5 (APA, 2013, pp. 761 ff.), we find an optional module for personality disorder diagnostics that presents a hybrid account of dimensional and categorical judgements of diagnostic features. It includes a Personality Functioning Scale with four dimensions (identity, self-direction, empathy, intimacy) on which patients may be rated on a scale from 0 (little or no impairment) to 3 (severe impairment) and a list of personality features to evaluate as present or absent. Suffi-

6 These changes were introduced following the increased interest in psychiatric research in thinking of at least some psychopathological features as occurring on a spectrum. Dimensional symptom ratings have been introduced as mandatory in the evaluation of diagnostic criteria for some mental disorder categorisations in the DSM-5 (e.g., autism spectrum disorder, intellectual disability) and as optional for others (e.g., primary psychotic disorder and personality disorders). Dimensional ratings have been made mandatory in some disorders categorised by ICD-11 (e.g., autism spectrum disorder, personality disorder) and optional for others (e.g., primary psychotic disorders). Also, they have been adopted in one way or another by relevant research movements in the field, such as the National Institute of Mental Health RDoC Project (NIMH, 2013) and the HiToP Research Consortium (Kotov et al., 2017). Here I will discuss only the case of personality disorders because my sole aim is to show how, in principle, my approach harmonises with this line of diagnostic. Discussing the scientific and clinical motivations for a dimensional understanding of mental disorders is beyond the scope of my project. For discussion of these motivations, see Krueger and Bezdjian, 2009; Helzer et al., 2009; Adam, 2013; Reed et al., 2019.

ciently high ratings on several scales plus the presence of relevant personality features may fulfil the symptom requirements of a syndromal diagnosis of a personality disorder such as schizoid personality disorder. The ICD, by contrast, offers an (almost) purely dimensional account. Like the DSM, it presents us with several domains tracking disturbances in functioning of aspects of the self and disturbances of interpersonal functioning that must each be evaluated for its pervasiveness and severity.[7] Although no explicit rating scale for evaluating these broader domains of self or interpersonal disturbances (like the one presented in the new DSM-5) is given, the new ICD contains a general scale that requires the clinician to judge the patient in their overall personality functioning as having a mild, moderate, or severe personality disorder. However, the new ICD approach to personality disorder diagnostics is only *almost* dimensional because it also contains specific features to add to the diagnosis, called "prominent personality traits or patterns". These denote striking features of personality disorders that previously were hallmark features for the categorical diagnosis of personality disorders. They include "borderline pattern", apparently akin to what previously was considered a borderline personality disorder, and "dissociality", apparently linked to traits previously thought of as specific to antisocial personality disorder.

In the end, the assessment of complaints that might suggest a potential evaluation as a psychopathologically relevant personality feature drawing on the newly introduced dimensional scales for symptoms is not very different from the evaluations discussed so far. For those symptoms whose presence is still intended to be evaluated categorically (character traits in DSM, prominent personality patterns in ICD), there is a clinical understanding of what constitutes these features on the level of the patient's behaviours, cognitions, and experiences such that complaints initially making the presence of this features a reasonable diagnostic possibility than can be evaluated against sets of propositions for the psychiatric symptom in question, as well as alternative models as discussed earlier in this chapter. When it comes to the dimensional assessment of symptoms, diagnostic practice can be best understood as operating such that each level of a symptom in question has an underlying set of propositions for the level of the system that is then intended to be evaluated against the patient, in addition to alternative sets of propositions that would render the complaint not a psychiatric symptom, but instead, for example, a distressing

7 The self-disturbance scale includes stability and coherence of one's sense of identity; ability to maintain an overall positive and stable sense of self-worth; accuracy of one's view of one's characteristics, strengths, limitations; and capacity for self-direction (ability to plan, choose, and implement appropriate goals). The interpersonal functioning group contains: interest in engaging in relationships with others; ability to understand and appreciate others' perspectives; ability to develop and maintain close and mutually satisfying relationships; and ability to manage conflict in relationships.

but not pathological psycho-behavioural occurrence. The DSM and ICD straightforwardly support the idea of evaluating the applicability of a set of propositions against the patient by providing short qualitative description of what features would have to be evaluated for different symptom levels.

If, for example, we consider a patient who reports that she has problems with "knowing what I want", the psychiatrist may include in the list of potential diagnostic evaluations of this complaint a symptom that is a feature of personality pathology, called "self-direction". Evaluating self-directions requires information on (1) the patient's goal-setting and goal-pursuing behaviour, (2) the qualities of the patient's setting and pursuing of normative standards for behaviour; and (3) the patient's capacity to reflect on an interpret the meaning of her own experience. If the individual shows Level 1 of impairment ("some impairment") on the sub-aspect of goal directedness, she is either "excessively goal-directed, somewhat goal-inhibited, or conflicted about goals" (APA, 2013, p. 775), whereas if she shows Level 2 of impairment ("moderate impairment"), "goals are more often means of gaining external approval than self-generated, and thus may lack coherence and/or stability" (ibid., p. 776). From this description of the three levels of the self-direction symptom, one can construe a set of propositions for each level that can then be used as a set of propositions to be evaluated against the patient's presentation in the in-depth evaluation, to support an evaluation of the complaint of "not knowing what one wants" as, for instance, moderate impairment of self-direction.

Similar descriptions can be found for sub-aspects of the descriptions of personality disorders of different severity in the ICD. If we focus solely on the aspect of interpersonal relationships, the description of "moderate personality disorder" (see Bach and First, 2018, Additional File 1) reads as follows:

> There are marked problems in most interpersonal relationships and the performance of most expected social and occupational roles are compromised to some degree. Relationships are likely to be characterized by conflict, avoidance, withdrawal, or extreme dependency (e.g., few friendships maintained, persistent conflict in work relationships and consequent occupational problems, romantic relationships characterized by serious disruption or inappropriate submissiveness).

Meanwhile, in the same symptom domain, someone with "severe personality disorder" is expected to show that their "problems in interpersonal functioning seriously affect virtually all relationships and the ability and willingness to perform expected social and occupational roles is absent or severely compromised" (ibid.).

Thus, if a psychiatrist is diagnosing a patient who reports that she has interpersonal problems that the psychiatrist also picks upon in the screening process, the psychiatrist will set up various sets of propositions to test, via the in-depth evaluation, whether the complaint might indicate a psychopathological problem. The psychiatrist will also derive sets of propositions from the qualitative descriptions just presented, and will likewise evaluate them in the context of the in-depth evaluation. The derived list of propositions for moderate interpersonal problems might then look like this:

- problems in most private social relationships
- problems in most professional social relationships
- few friendships maintained
- persistent conflict in work relationships and consequent occupational problems
- romantic relationships characterised by serious disruption or inappropriate submissiveness

To give a brief idea of something that might also be considered by a psychiatrist in the case of a complaint regarding recurring interpersonal problems, an alternative evaluation of the complaint of repeated interpersonal problems might be that the patient suffers experiences of repeated social exclusion for other reasons. To take an example from my own clinical work, the patient may suffer from a hearing impairment that leads him to misunderstand or miss what people say if he is not fully concentrated on the conversation, and he may feel bad about his problem and so not tell anyone about it. The result is communication problems that may be misunderstood as ignorance or just weirdness on his part, which leads people to withdraw from him. If we want to put this in a list of features this can look as follows:

- The patient has a physical impairment that complicates communication.
- The patient does not usually speak openly with others, or even actively hides the impairment from them.
- People tend to retreat from social contact with the patient, saying they feel ignored by the patient or that the patient forgets things they have said.

The DSM and the ICD descriptions of the prerequisites for dimensional categorisation in these domains of personality functioning both offer propositional descriptions that differ from each other either in the extent to which a problem seems to be present (as in the ICD example) or in the quality of the phenomenon rather than only in quantity (as in the DSM example). These descriptions can be used as sets of propositions to evaluate the level of the symptom via the in-depth evaluation of the recognised complaints that might indicate these psychopathological problems.

So far, I have discussed clinical examples of the complaints of reluctant speech and constant worrying and have presented some potential clinical evaluations that a psychiatrist may consider if a patient presents with this complaint. I have also spelled out some of what a psychiatrist would look for to support these potential diagnostic evaluations of complaints. In addition, I took a moment to discuss how, as part of this way of describing the diagnostic procedure, we can understand the relatively new approach of evaluating symptoms by level of severity. Here I outlined the screening step of the diagnostic procedure and the reasoning that takes place in preparation for the in-depth evaluation. Now let me come to the task of mapping part of the criteria for something being a qualitative, constitutive diagnostic modelling process (which I presented in the introduction to this chapter) onto the clinical process illustrated here.

3.1.2 Mapping Modelling onto Screening

Let me now turn from this description of the initial clinical reasoning process, illustrated with specific examples, to show what aspects of diagnostic modelling are embodied by it. I begin by considering the initial screening of the patient, and with it the psychiatrist's initial recognition of their complaints.

As described earlier, the initial screening involves letting the patient report her reasons for wanting to speak with clinician, as well systematically exploring aspects of the patient's experience, including their psychological as well as behavioural functioning, by questioning and observing them. Thus the complaints of the patient are identified. Complaints are abnormalities in the assessed aspects of the patient that might indicate the presence of a psychopathological condition. This initial step of the identification of complaints equals the step of diagnostic modelling that I called *prima facie* error recognition. In this initial step of diagnostic modelling, the diagnostic modeller flags outputs of a system that might indicate an error in the system. This recognition in modelling takes place based on what I called the normative model of the system. This is a model that indicates which kinds of output should be expected in a well-functioning system under the usual conditions in which the system operates. If developed in detail, the model also fleshes out some details about the inner processes of the system associated with normal system outputs.[8] Based on their ideas about how human experience and behaviour, if not potentially psychopathological, is supposed to appear in individuals in what one may consider a rage of normal life circumstances, the psychiatrist (just like the diagnostic modeller) will note the deviations from the assumed range of normality and suspect that these might indicate the presence of a psychopathological symptom that in turn indicates a psychopathological condition in the patient – or an error in the system, as the diagnostic modeller would say. We may think of the background assumptions of the psychiatrist as a long list of features that should be considered within the scope of typical human psycho-behavioural phenomena, not potentially indicating an instance of psychopathology. More specifically, one plausible way to think about these background assumptions made by the psychiatrist is as a large set of propositions

8 For a more detailed articulation, please see Chapter 2.

representing what is assumed to be potentially normal. Such propositions might include: People can fall asleep after being awake for a workday. People have things that bring pleasure to them. People are in in control of their thoughts. If the psychiatrist picks up information about the patient indicating that one of these propositions does not apply to them, this indicates a complaint (in this case, for instance, sleeping problems, loss of interest, or intrusive cognitions) that will be considered for the in-depth evaluation. If we consider such sets of propositions to be qualitative models that are meant to represent a normative state of the system expressed in terms of propositions (which I will make more plausible when I talk about propositional models below), this set of normative assumptions that psychiatrists have in the back of their mind while talking to patients would plausibly qualify as a normative model and the complaints would then plausibly match up with the suspicion of an error in the system evoked by the recognition of a *prima facie* error, which will then guide further diagnostic efforts in diagnostic modelling.[9] This seems to establish the fulfilment of the initially presented criterion (*v*) that there is a normative model at work in the initial recognition of a *prima facie* error.

As we saw above, initially recognised complaints then drive suspicions about potential situations that might have led to their presentation. These suspicions take the form of diagnostic hypotheses that might lead to various evaluative outcomes. The complaint might be evaluated as an actual medical problem that then might either fall either into the realm of psychiatry or be categorised as a medical but non psychiatric problem. Or the complaint might be evaluated as not being a medical problem at all. These hypotheses, as we saw, come with a set of propositions whose evaluation is used to enable decisions about which diagnostic conclusion should be drawn. But before I come to the process of diagnostic decision-making, let me give a little more time to the advancement of the diagnostic hypotheses and their subordinated sets of propositions, in connection with my understanding of modelling.

At the beginning of the step of diagnostics that I have just discussed, namely screening, the psychiatrist puts forwards multiple ideas as to what might be the patient's problem. These proposals have diagnostic labels (psychopathological condition X, medical non-psychopathological condition Y, or a type of non-medically relevant complaint) that are accompanied by theoretical structures consisting of sets

9 Note that the term *normative model* applied in this situation is theory-neutral insofar as it remains silent about what are or should be the sources of such normative standards. In this way, I can avoid engaging in the ongoing debate between those who consider our understanding of psychopathology to be best analysed in terms of natural functions and those who believe that we must consider normative judgements to feed into our understanding of what counts as mentally healthy or not (on this debate, see, e.g., Faucher and Forest, 2021). Both assumptions are compatible with the idea of the use of a normative model in psychiatric diagnostics; all that would change is what the final justification of such normative assumptions about a well-functioning system would be. The answer to this question does not affect my account.

of interrelated propositions hypothesising certain states to be present (or to have occurred) in the patient and how these states relate to each other or to other states in the patient. These sets of propositions are intended to match aspects of the patient under diagnostic evaluation. The initial diagnostic ideas and their propositionally structured package are put forward by the psychiatrist based on little initial information about the actual patient at hand, usually just the report of an experience offered by the patient or an observation made by the psychiatrist that is in no sense treated as sufficient to establish a diagnostic conclusion right away. The sets of propositions related to each diagnostic label considered to possibly apply to the patient, rather than being informed by substantial information about the patient at hand, is informed by the psychiatrist's scientific and clinical background knowledge as well as their common-sense psychological understanding of the human mind.

Considering this process so far, it seems that the sets of interrelated propositions that are intended to fulfil representational functions in relation to a real-world system (the patient) present a structure that would qualify as a candidate for a propositional model structure (Thomson-Jones, 2012). As discussed in the last chapter, propositional models are not a model in the sense most often used in science, where models are mostly specified as quantitative mathematical structures. Nor is this type of model specified in one of the more formal ways offered by qualitative mathematics, or in terms of a box-and-arrow graphic with a legend that assigns meaning to the components of its structure. Rather, propositional models are qualitative model structures that consist of propositions whose content expresses a state of affairs meant to apply to the real-world system that the model targets. If we stay close to clinical reality – that is, a clinician who thinks about what would have to apply to a patient for their complaint to constitute a certain symptom, which will lead him to bring to mind what has to be true about the patient to have this symptom – it seems natural and *prima facie* most plausible to think of what comes to his mind as a set of language-like propositions presenting a list. Just like the ones I introduced in the last subsection, this list sums up the different facets of what should be true about the patient in order to provide this or that clinical evaluation of a symptom. Thus, when psychiatrists do employ qualitative models in diagnostic reasoning, those models seem to be best understood as models consisting of sets of propositions – that is, as propositional models. Whether these supposedly propositional models really are models, as defined in the approach being presented here, then of course depends on whether these structures are set up and handled in accordance with the indirect strategy of representation. If this turned out to be the case, then PD would meet criterion iv. To take the first step in showing that the whole process indeed qualifies as modelling, let us now turn to how the representational structures used in diagnostics are set up.

To determine whether the theoretical structures used by the psychiatrist to identify complaints and evaluate them diagnostically qualify as model structures we need to assess whether they are set up and used in the three-step procedure of model construal, model analysis, and model/world comparison, as discussed in detail in the last chapter and as also set out as criteria (*i*), (*ii*), and (*iii*) at the beginning of this chapter. Let us first look at model construal.

The theoretical structures that seem to be used by psychiatrists to target their patients seem to be set up based on little initial background knowledge about the actual system intended to be represented by the regarding structures. Instead, the structure itself is provided from a canon of background theories and the reuse of models from psychopathology, medicine, and common-sense psychology for the conditions that might be present in the patient. The theoretical structures used by the psychiatrist to identify complaints and evaluate their diagnostic status are set up based on various sources of inspiration, but the previous direct investigation of the system makes the process of setting up the structure equivalent to what, in Chapter 2, I discussed as constituting *model construal*. In other words, as discussed, the modeller does not start with an investigation of the modelled system and derive their model from the investigation, but instead brings to the table a pre-established idea of the structure that will be used to represent the modelled system. The psychiatrist has in mind a pre-established understanding of what constitutes complaints and the presence of specific diagnostic evaluations of these complaints, and does not develop such ideas anew when engaging with every single patient. The psychiatrist has a pre-established understanding of psychopathology that he can recall in the form of sets of propositions. It seems, therefore, that the first important point, and criterion (*i*), the model construal, is met. Next up in the process of modelling would be the *model analysis*.

Model analysis is the step in which a modeller considers the implications of the model structure that has been set up. These include which aspects of the real-world system are meant to capture which aspects of the real-world system (*model scope*); which aspects of the model are assigned to which specific parts of a real-world system at hand (*model assignment*); and how well a model's elements and relationships amongst them, including the impact that a change in one part of the model should have on the rest of the model, matches the real-world system's makeup and behaviour (*representational and dynamic fidelity criteria*). The model analysis step is typically explicitly present in modelling only if a model is being set up for the first time or is being undertaken more thoroughly than usual by someone using a pre-established model or using a model derived from a theory when this model is newly learned. An experienced modeller who commonly uses one and the same model structure multiple times will not need to analyse the model every time it is used, because they already know its implications. This is also the case with diagnostic experts.

Think of a psychiatrist who is well trained in the theoretical understanding and clinical appearance of the symptoms of anhedonia. He does not have to think through his knowledge of anhedonia to become aware of the model's assumptions and its implications every time he considers a patient to suffer from this symptom. He will know them by heart. Such well-established knowledge of a clinician about any symptom will encompass the different aspects of model analysis – setting the scope of the model, model assignment, and providing representational and dynamic fidelity criteria – as such knowledge about symptoms entails a good understanding of propositions, describing states or dispositions whose absence or presence are relevant to come to diagnostic decisions regarding the presence of complaints and symptoms. This understanding includes which features of patients are relevant to look at (*scope of the model*) and which propositions are referring to which features of the patient (*model assignment*) – which is rather self-suggesting by the proposition's content (i.e., the meaning of the words in the proposition). All that is needed is an adequate understanding of these meanings of the content of the propositions. Take, for example, a diagnostic proposition that the patient wakes up at night in terror because of bad dreams. It would be clear that the scope of this proposition, which makes its target part of the scope of the overall model, would by virtue of the proposition's content be the patient's sleep behaviour and dream experiences, and also that the aspect of model assignment for this proposition as part of the model would be taken care of by its meaning – namely, the target of this specific proposition would be the patient's sleep behaviour and dream experiences.[10] Now let me come to the last aspect of model analysis setup: fidelity criteria.

The thorough understanding of psychopathology that diagnostic experts such as well-trained psychiatrists bring to the table also takes care of the last aspect of model analysis, fidelity criteria. To recall, fidelity criteria are the criteria for how well a given model structure (in our case this would be a set of propositions) is supposed to map onto the elements or processes of the model's real-world target in order to consider the model permissible. Showing that diagnostic reasoning preparing the in-depth evaluation also sets up fidelity criteria requires a bit more discussion. This discussion is required thanks to the role played by vagueness in this context, which may initially provoke doubt if indeed fidelity criteria are generally assumed. If not, this would undermine my claim that model analysis takes place, and that this portion of diagnostic reasoning is modelling at all, thus endangering my whole project. Therefore, I will argue that what we see in diagnostic reasoning that employs natural language propositions is the occasional vagueness that we encounter in language on a regular basis, which does not mean that fidelity criteria are absent; they are just

10 A mathematical model, on the other hand, consisting only of quantitative constants and variables would not be so straightforwardly interpreted. Language-like propositions have a content whose meaning is indicative of its target, while for numbers and symbols being mapped onto a feature of a system, either widely known conventions or intentional assignments are needed.

vague sometimes. As is often the case with qualitative theorising, cut-offs for when a qualitative representational structure such as natural language propositions map onto its target have vaguer boundaries than quantitative representational structures employing numerical values that can be mapped onto numerical measurement outcomes of real-world systems.

In the case of the theoretical structures, vagueness comes into psychiatric diagnostics on the level of diagnostic propositions. First, propositions used to evaluate potential diagnostic interpretations of the patient's presentation are in themselves vague, and second, it is vague how many of these propositions must apply to justify the assumption that the theoretical structure maps onto the target. Let me discuss both aspects. Regarding the vagueness of diagnostic propositions, we may notice that they often contain vague phrases. By this I mean phrases that by virtue of their meaning do not provide a clear-cut criterion for when they should be applied, but leave room for borderline cases. Borderline cases in normal language use are, for example, the use of the word "dusk", where it is hard to say when exactly it begins or ends, or the correct application of the phrase "heap of sand", when we look at a growing collection of grains of sand asking ourselves how many grains are needed to make a heap. In diagnostic propositions we do not talk about dusk or sand but sometimes, as in the set of propositions for pathological generalised worry discussed earlier, similar vagueness creeps in. We read that worry is "difficult or impossible to dispel" rather than easy to dispel. But when exactly does it become difficult rather than easy to dispel a worry? It seems that in attempting to pin down the meaning of "hard-to- dispel worry", we cannot provide a definite answer. Or consider another diagnostic proposition saying that worries are supposed to be "closely" linked to current concerns. How close is closely, and when does the worry start to be linked distally or semi-closely? Or, to take another example that will be discussed below in detail, look at a part of the ICD-11 (WHO, 2019) criteria for severe personality disorder: "problems in interpersonal functioning seriously affect virtually all relationships". Again, when precisely do the effects start to be "serious", and how many of the relationships must be affected to count as "virtually all"? There might be clear cases in which we would say that it is hard rather than easy for someone to dispel a worry, that a worry is closely linked to current occurrences rather than only distally linked, and that virtually all rather than only many of a person's relationships are affected, but there may also be cases where we struggle to draw the line between these alternative evaluations. The vagueness of the applicability of single diagnostic propositions propagates to the set of propositions containing them. If vagueness can make it challenging to determine whether one proposition of a set of propositions apply to a patient, the same will be true when the task is to decide whether the set of propositions applies to a patient if one part of the set is a proposition that actualised the problem of vagueness in a concrete case that the set should be applied to. Therefore, although diagnostic propositions can be considered to inform a diagnostic expert about their fidelity criteria by virtue of their meaning, the fidelity criteria via which they do are in themselves not clear cut.

However, vagueness of criteria is not the same as absence of criteria. Although we encounter vagueness in the employment of diagnostic propositions by clinicians, this does not mean that clinicians assess the presence of psychopathological conditions in their patients with no idea when a proposition matches with the patient's presentation and when not. It is simply the case that in some instances, it will be not straightforward to decide this question, and these cases are borderline cases. In these cases, just as in the case of model scope and model assignment, fidelity criteria are present, in the form of the meaning of the propositions, but when exactly the proposition applies will on occasion be undecidable due to the vagueness of these meanings.[11]

A further point worth noting about fidelity criteria is that, just as in any other case of modelling, the purpose of fidelity criteria is to say how good the match between model and world must be in order to accept the model as a model of the real-world system, and this purpose allows for some error. We also see this indicated explicitly in sets of diagnostic propositions. This occurs most obviously in cases where diagnostic propositions themselves contain phrases like "usually", "often", or "regularly". Again, in the case of pathological generalised worry, we might say that these worries are supposed to "predominantly take the form of verbal reasoning". That means that there will also be cases where these pathological worries are not verbal. So, it seems that this proposition, although likely true, does not *have* to be true in order for a psychiatrist to apply the label of pathological worry that is provided based on the match between diagnosis proposition and a patient. On other occasions, this room for error may not be directly expressed in the proposition (though it could be) but will be considered by the psychiatrist based on common background knowledge.

---

11 It is worth noting that occasionally a psychiatrist will make decisions regarding borderline cases influenced by non-theoretical factors, practising higher-order reasoning in the clinical context. To explore these practical rationales systematically is beyond the scope of this chapter and would strictly speaking no longer fall under diagnostic reasoning proper, rather addressing an impact of practical clinical reasoning on diagnostic reasoning. Think, for example, of the potential positive and negative consequences that the decision may have for the patient. If, in the end, ascribing this proposition would lead to the ascription of a symptom that would lead to a diagnosis that would in turn have a serious negative impact on the patient's life – for example, lead to the prescription of medication with severe side-effects or impact the patient's ability to work in certain sectors – these complications may make the clinician who has the overall good of their patient in mind hesitate to count borderline cases as positives and inclined to make conservative diagnostic decisions. The same may apply the other way around if the potential prescription of medication that might save the patient's life, and whose prescription would have no serious side-effects, depended on a positive evaluation of a borderline case. In these cases, many clinicians will find themselves inclined to be more liberal in their judgements.

If, for example, we look at traumatic brain injuries (TBI) discussed earlier (section 4.1.2), they contain the criterion that lesions that indicate traumatic brain damage can be seen in computed tomography. If one digs into the relevant literature (see the original discussion of TBI for references), it is clear that if enough of the other expected aspects of TBI are present, clinicians are nonetheless willing to diagnose TBI even if there is no lesion on the CT.

It seems that each set of diagnostic propositions (or propositional diagnostic models, as I suggest we can consider them) also has its fidelity criteria for which and perhaps how many propositions must apply contingently or necessarily to make the propositional model map well enough onto the patient presentation to make the model acceptable.

We know the requirement that some core features must apply, and then other features may apply. However, in difference to higher levels of diagnostic decision making (the formalised criteria for what symptoms must be present in order to diagnose a certain disorder according to DSM or ICD) on the level of symptom diagnostics we are currently looking at there is no official standardised manual. Instead, the clinician's psychopathological, medical, and commonsensical psychological understanding of the specific potential diagnosis they are evaluating will influence how well they expect the model of, for instance, pathological generalised worry to map onto the patient reporting constant worrying, and whether they infer that it is matched well enough to accept this model for the patient. This might even result in differences on the level of diagnostic decision-making regarding the attribution of symptoms, depending on how up-to-date the clinician's understanding of the relevant condition is. This is a topic that will be explored in more depth in the next chapter when I discuss diagnostic errors and disagreements.

Considering the discussions in the preceding paragraphs, we find that there are fidelity criteria, though they are occasionally vague, for the sets of propositions used for diagnostic purposes. This means that all steps of model analysis (scope, assignment, fidelity criteria) take place in diagnostic reasoning as it can be expected to proceed in cases in which well-known models (here, sets of diagnostic propositions) are reused by experienced modellers (psychiatrists), such that criterion *(ii) model analysis* can be considered fulfilled. Now let me come to the last criterion, criterion *(iv) constitutive models*, that I will show to apply to the screening procedure.

As a last point in this section, I want to show that the procedure described above employs theoretical structures that, assuming they are models, would meet the requirements of the initially introduced criterion *(iv)*, and thus can be considered constitutive models. As discussed in detail in the last chapter, to qualify as constitutive, models must point out factors of the system that they attempt to represent that, if present, would be components of the system providing it with the power (or disposition) to bring about the phenomenon that the diagnostic model is intended to indicate. Does this apply to the sets of propositions used in psychiatric diagnostics if we consider them to be models? I have two reasons to think so. One is a *prima facie* reason, making this option more plausible; the second is a reason that can be demonstrated via examples, as I will do by coming back to the case of pathological generalised worry. Let me begin with the *prima facie* reason.

The *prima facie* reason to believe that if diagnostic reasoning is modelling, it is constitutive modelling is that there is a constraint on what kinds of modelling it could be, and among the alternatives, constitutive models seem to be the most plausibly attainable in clinical diagnostic contexts. Let me elaborate. As briefly mentioned in the Introduction and discussed in detail in Chapter 1, it seems that models used for diagnostic purposes must do explanatory work, since it is required of the diagnostic process that it produce a case formulation equipped with the capacity to explain the patient's condition. If this case formulation is derived from the earlier diagnostic process, and if we accept that this process is a modelling process (as I argue it is), the explanatory power of the case formulation must be generated by the diagnostic modelling process that provides the material for the formulation. This in turn means that the models from which the formulation would derive must do explanatory work. If we commit to the requirement of explanation and modelling and thus locate the source of explanatory power in the modelling process informing the diagnostic case formulation, two relevant class of models discussed in detail in the last chapter are potentially option: explanatory modelling may be either causal or constitutive. Let me give a quick reminder of what we are talking about when we talk about causal or constitutive models.

Causal models would aim to point out the causal aetiology – that is, the chain of events and its stages – of the occurrence of an output of the system, such as the complaint of the client. A causal model explaining a type of symptom to be present in a system would have to explain it as a causal consequence of a specific kind (or class) of causal process; it would say *why* the system is doing what it is doing. To then diagnose a condition with such a model, we would need to evaluate whether the causal story that the model tells us is in place within the patient. Using a constitutive model, by contrast, allows us to dispense with looking at the exact causal processes in the system because, to quote Cummins (2000, p. 122) again, such a model "abstract[s] away from the behavior and orchestrated activities of the parts and ask[s] how the system has a capacity for this kind of behavior". Abstracting away from causal details in this way makes things easier. It will ostensibly often be simpler to identify reliable constituents of a system on some level of description that our investigation tells us is responsible for an output, and then to assess the presence or absence of these features, than it is to come up with a detailed explanation entailing all these components, plus a story about how their interactions produce the output in question, and then to assess whether precisely this process has taken place. Finding reliable causal explanations for phenomena is a notoriously complicated task, especially in complex systems like human minds and behaviours, and it is usually easier to merely identify the components that presumably put the system in the position of producing the output, without getting into the details of why they do this.

Considering these two forms of explanatory modelling, a brief look at psychiatric diagnostics makes it implausible that, if it is modelling, what is being used in it are causal models. It is implausible because psychopathology has produced virtually no such models that could be used for diagnostics, nor do the methods of assessment in psychiatry seem to be suited to evaluate causal claims. Let me elaborate. Causal models of how specific symptoms emerge that are widely accepted as the basis of a psychopathological understanding of specific symptoms that is also used as the basis of clinical assessments, and that track down the relevant causal process that can be assumed to generally occur in patients if they suffer from a symptom, are beyond the current reach of psychiatric science. Considering the current state of our psychopathological understanding, we arguably have no widely accepted model for psychiatric symptoms of major psychiatric disorders that allows us to understand the causal process producing it and that could be used in clinical practice. There is, for example, no causal model of hearing voices in schizophrenia that is widely accepted in psychopathology, that is so reliable that it is used to assess whether a patient reporting hearing voices even though no one is present is suffering from verbal hallucinations. Such models may be developed and used in the future, but they are not part of clinical reality at present.

Moreover, commonly used diagnostic procedures at the core of diagnostic practice seem to carry out a comparison between a causal model and the presentation of the patient that would allow us to infer causal relationships. In psychiatric interviewing, the mental status examination, or the commonly used cognitive tests as discussed in the examples in this section, it does not seem that what is being assessed in the models are either a) counterfactual relationships amongst the elements (i.e., propositions) of the models, as would be required by a counterfactual approach to causality (Menzies and Beebee, 2020), or b) alternative criteria commonly treated as indicating causality, such as the Bradford Hill criteria (Hill, 1965), which track down plausible causal pathways or identify strengths of association between supposedly causally interacting elements (e.g., dose–response relationships). Nor do the models ensure what has more recently (Cartwright, 2022) been claimed to be required to support causal claims in single cases, like the "elimination of alternatives" (ibid.) ensuring that no sources of bias are present (i.e., proclaim and control causally biasing variables).

One might suggest an objection at this point. The sets of propositions I have provided may indeed often point to features that would plausibly also play a role in the causal story of how the complaint occurs in the diagnosed patient. However, setting up a model of a system that is representing crucial features of a system that contribute to an output of the system, such that by the application of the model the presence of these features is evaluated, is something different from setting up a model of the causal process of which the features of the system addressed in the model are part. We might start with the simple fact that the latter case will contain a claim about some elements causally influencing other elements, while the earlier model, addressing features alone, will not make claims about such causal dynamics amongst model elements. A diagnostic proposition – for example, the one indicating a blow to the head with loss of consciousness and other mental complaints assessed in the context of the assessment of reluctant speech being due to TBI – certainly assesses a feature of the system (i.e., having been in a certain state due to a specified occurrence) that plausibly may also have played a role in a causal story of the psychological complaint, if this complaint is indeed connected to a case of TBI. But this proposition makes no claim about the occurrence causing the reluctant speech, nor is there any mention of how this proposition is supposed to be linked to the other propositions in the model to indicate causal relationships between them. Of course, it would be highly plausible that, to pick out another diagnostic proposition, a brain lesion found in a CT might be the result of an impact to the head, and we might be very likely to consider this to be the case if both propositions apply. However, the model itself does not establish this claim or provide guidance to assess any causal relationship between a potential blow to the head plus its immediate psychological consequences and the finding of a brain lesion. It just asks us to evaluate whether the patient has experienced such a blow to the head and/or has a brain lesion; it does not engage in causal claims.

What to make of this? If the propositional models used in psychiatry are perhaps not causal models, given the lack of a good causal understanding of psychiatric symptoms in psychopathology, and the apparent fact that the diagnostic evaluations are not tracking down information suited to evaluating causation (although they do identify features of the system that play a role in the system executing the causal capacity to produce the complaint) this should ring a bell: These features might instead be constitutive factors. However, to support the claim with more than a plausibility argument, let me present my second reason.

The second reason why I argue that the sets of propositions used by psychiatrists, if they are models, qualify as constitutive models derives from how they are best understood to account for the evaluation of a complaint. To illustrate this, I will look again at the set of propositions used to evaluate the presence of pathological generalised worry. Here again is the set of propositions proposed to be used to evaluate this condition:

- predominantly takes the form of verbal reasoning
- is non-specific, abstract, or general in content (e.g., "what if *the worst* happens?")
- is persistent (i.e., of long duration)
- is closely linked to the individual's current concerns

- is attributed to an internal origin (i.e., the patient assumes ownership of these thoughts)
- is experienced as egosyntonic
- is difficult or impossible to control or dispel
- entails a stress-inducing faulty appraisal concerning whether the feared consequences might come to pass ("worry about worry")
- is thought to have the positive power to potentially prevent the feared event.

If we think of this set of propositions as a model, this model consists of nine propositions. Taken as a whole, the model presents criteria to be met by a patient experiencing constant worries in order for these worries to be evaluated as suffering from pathological generalised worry (PGW). In other words, if we think of the experience of constant worry as an executed disposition of the system producing them, rather than just an occurrence, this propositional model points out factors that should be true of the system, actualising or executing the disposition to constantly worry, in order to justify the evaluation of the worrying as PGW. Again, it does not seem to provide a causal account as to why these worries occur. What instead makes the features pointed out by the proposition's constitutive factors – that is, features that justify us in saying that the executed disposition is PGW – is the idea within the model that it is this feature of the worrying system that makes this worry be the executed disposition of PGW. To come back to the example from the last chapter, what makes something have the disposition to be fragile is (given some background conditions) that it breaks when falling from hip height, so that the feature of breaking when falling from hip height is a feature of glass that constitutes its fragility. Similarly, it is experiencing worries as egosyntonic, the worries being predominantly in the form of verbal reasoning (and so on), that makes the occurrence of constant worry the actualised disposition of PGW. In this way, these features are supposed to account for the instances of worrying as being the execution of the disposition to PGW, just as the feature of breaking when being dropped from hip height is what makes something have the attributed disposition of fragility, which is executed when dropped. Thus, what is pointed out by the model and therefore looked for in the patient's evaluation are constituents of their psychopathological state, thought of as dispositions. These dispositions will thereby be explained by providing the constitutive features that are relevant to make the difference between evaluating the system as having this disposition and (if these features were absent) not having this disposition.[12]

12 You might note that the features pointed out by diagnostic propositions are widely different things. Some features of what a disposition requires for its attribution may involve, for example, something having happened to the system in the past; this can therefore be a constituent of the system having this disposition. Imagine there were the disposition to go to heaven after death and we had a word for it. If to have this disposition a human had to be touched by a

The same could be shown for the other examples of collections of propositions employed for the purpose of diagnostic assessment that I discussed in 3.1.1. It therefore appears that if we consider these sets of propositions – set up as the outcomes of the screening procedure to be used for the in-depth evaluation – to be propositional models, these models, which have to do explanatory work to support the subsequent diagnostic case formulation, are constitutive (rather than causal) in nature. Hence, criterion (*vi*), which requires that the models used in psychiatric diagnostics be constitutive, is fulfilled. And thus, all criteria I intended to show to apply to this stage of the diagnostic process apply.

In this section I began to argue that a large part of the diagnostic process can be understood as qualitative, constitutive diagnostic modelling. To do so, I showed how several of the criteria I set out in the introduction to this chapter that must match with the diagnostic process do indeed apply to the diagnostic screening process and the preparations made within it for the in-depth evaluation. More precisely, I argued that the screening procedure and the preparatory steps for the in-depth evaluation can be matched with criteria (*i*), (*ii*), (*iv*), (*v*), and (*vi*). I showed that if we think of the theoretical structures used by psychiatrists in their reasoning as models, the initial error recognition can be thought to take place *qua* normative models (*v*); that if we think of theoretical structures employed by psychiatrists in the context of diagnostic reasoning as models they should be assumed to be qualitative, more precisely propositional models (*iv*); that if we think of them as models, they should most plausibly considered to be constitutive models (*vi*); and that the way the psychiatrists derive and think about the theoretical structures used in the context of diagnostic reasoning corresponds to what we would expect of model construal (*i*) and model analysis (*ii*). To show that diagnostic reasoning can indeed be understood in large part as modelling, only one more criterion must be demonstrated to apply – namely, that a model/world comparison (criterion (*iii*)) takes place, and more specifically that diagnostic reasoning is the kind of diagnostic modelling I presented in the last chapter. I then still need to show that it employs model-based exclusion and differential diagnostic (criterion (*vii*)). Presenting arguments in support of both criteria will be the task of the next section.

---

holy person, at least one constituent for having the disposition of going to heaven would be something that occurred to the system in the past. There is no problem with considering such instances and other features of the system that are not physical parts of the system here and now to be constituents for it having a disposition. Thus, it is no problem that diagnostic sets of propositions meant to enable diagnostic evaluation of complaints as types of dispositions contain propositions referring to such features.

3.2 Diagnostic In-Depth Evaluation as Modelling

The last section focused on the screening procedure in psychiatric diagnostics and how it sets up the next stage, the in-depth evaluation. This section will focus on in-depth evaluation and how the criteria set out in the introduction of this chapter are realised by it. I will argue that in the in-depth evaluation we see the model/world comparison take place (criterion *(iii)*) and that we observe the occurrence of model-based differential and exclusion diagnostics (criterion *(vii)*). By demonstrating this, I will show that the screening and in depth-evaluation together meet all seven criteria I set out to be required to support the claim that this portion of the diagnostic process can be understood as operating via the method of qualitative, constitutive diagnostic modelling.

As I did in the last section, I will begin this section by discussing the progression of a potential clinical evaluation, based on the setup I provided in the last section. After a short recap of what in-depth evaluation is all about, I will (3.2.1) continue to use examples to illustrate this step of the diagnostic process. After this illustration, I will then (3.2.2) argue how criteria *(iii)* and *(vii)* apply to this part of the diagnostic process to drive home my point that screening and in-depth evaluation taken together are the portion of diagnostics that can be explained as following the method of diagnostic modelling discussed in the last chapter.

3.2.1 Example of In-Depth Evaluation

With the end of the systematic screening[13] procedure providing a list of patient's complaints and, based on these complaints, a list of several diagnostic evaluations of the complaints, the psychiatrist arrives at a set of diagnostic propositions for each of the diagnostic options for evaluating the complaint. Deciding which of the potential diagnostic evaluations should be selected to classify the complaint in terms of

13 Systematicity here means that the psychiatrist is not only considering spontaneous reports of the patient but also asks about unmentioned potential complaints that, if present, would also require more careful diagnostic attention. If, for example, in the course of this evaluation, the psychiatrist asks about the patient's relationship to his own body, or his body awareness, the patient may say that he often feels as if he were not in his body but "standing behind myself" or "not really looking though my eyes but like through swimming googles" – experiences that appear normal to the patient because, as he reports, he has had them from time to time for as long as he can remember. The description, however, suggest an anomaly that on closer investigation might turn out to be a mental symptom, namely *depersonalisation*, and the psychiatrist will include it in the list of complaints and come up with models that might apply to the patient's case, which in turn enable him to determine whether this complaint indeed is this mental symptom.

its psychopathological relevance is the job of the following step, the in depth-evaluation. In the in-depth evaluation, the psychiatrist evaluates the sets of propositions that make up the relevant understanding of what it is to have a psychiatric symptom, medical symptom, or non-medical complaint against the patient's presentation to select the corresponding set of propositions and, in accordance with this, the appropriate diagnostic categorisation. Let us look at this process in more detail.

In the in-depth evaluation, the psychiatrist is guided by the sets of propositions that have been selected in accordance with their ability to support potential evaluations of these complaints as psychopathological symptoms. The sets of propositions relevant to evaluating each complaint suggest what information would be necessary to support or refute the applicability of each proposition to the patient's presentation. The psychiatrist therefore begins a non-random but rather guided process to generate a set of information about the patient, specifically geared towards assessing the applicability of diagnostic propositions that require evaluation. The main means are, as discussed in the first chapter, interviewing and behavioural observation of the patient (MSE, psychiatric interview) and potentially also information gathered from their relatives, as well as potential further cognitive and biological testing. The information generated in these ways is collected and organised, be it only in the psychiatrist's mind or on their notepad, to present what in the last chapter I called a *prepared description* of the targeted system allowing for the evaluation of the set of diagnostic propositions.

Once the relevant diagnostic information has been collected and ordered, the preselected sets of diagnostic propositions can be compared with the collected diagnostic information for the specific purpose of evaluating which of these propositions apply to the patient at hand. The basic idea is then that this competence enables a decision of which (if any) of the sets of propositions apply, so that the diagnostic evaluation belonging to this set of propositions can be inferred. If the relevant set of diagnostic propositions is similar enough to the presentation of the patient – that is, if the fidelity criteria determining how good the match between model and real-world target must be are met – the evaluation will be accepted; if not, it will be rejected.

As discussed in the last section, to determine whether a proposition applies or not, and whether enough of the propositions apply (given that not all of them need to apply in any given case), is a task that may be impacted by the vagueness of fidelity criteria giving rise to borderline cases. This vagueness leave room for diagnostic judgement that must be exercised in the face of irreducible uncertainty in borderline cases. Let us look at this whole process again considering our examples.

Consider once more the patient's complaint of uncommon speech production, as observed by the psychiatrist in the context of the initial screening situation. Various potential states of affairs might be responsible for this presentation of the patient. Three of them were introduced: one where it would turn out that the patient is not suffering from any medically relevant symptom, either psychiatric or otherwise; one (traumatic brain injury; TBI) where the complaint would prove to be a medical-neurological problem; and finally, one (alogia) that would turn out to be a psychiatric symptom. To evaluate the first option, I considered the following set of propositions:

- The patient can elaborate their answers if asked to.
- The patient recognises that their answers are unusually short and can justify their manner of speaking by explaining their motivation (e.g., wanting to save the doctor's time, or wanting to be as precise as possible).
- The patient can report that his way of presenting information is not something that has developed recently but is rather their normal way of conveying information.
- People who know the patient report that the patient has always tended to speak this way.

To collect the necessary information about the patient to evaluate all the propositions that would support the evaluation of the complaint of reluctant speech according to this set of propositions and thus as an idiosyncratic and non-pathological manner of speaking, the psychiatrist can straightforwardly find out all he needs to know in conversation with the patient. She can simply ask the patient to elaborate an answer (first diagnostic proposition); can make the fact that the answers provided by the patient are rather short itself a topic in the interview and inquire into whether the patient is aware of the shortness of his answers and whether he is intentionally providing them in such a telegraphic style (second diagnostic proposition); can ask whether this way of speaking is typical for the patient or not (third diagnostic proposition); and may support the certainty of this last point by also speaking to people who know the patient better – likely to a family member (fourth diagnostic proposition).

If we alternatively consider the evaluation of speaking behaviour as not normal and also not a psychiatric symptom, but as the psychological side-effect of the non-psychiatric medical condition of TBI, the following propositions would have to hold true:

- The patient recently took some sort of blow to the head (e.g., by falling or having an accident).
- The patient did suffer some such blow, and lost consciousness or had loss of memory of events immediately before or after the blow.
- There were alterations of mental states at the time of the accident (e.g., feeling dazed, disoriented, or confused).
- Lesions that indicate traumatic brain damage be seen in computed tomography.

Again, the psychiatrist may gather the information relevant to fuel the prepared description of the patient by asking the patient questions to find out whether the patient recently took a blow to the head (e.g., in an accident) or by consulting medical records regarding potential accidents shortly before the patient saw the psychiatrist. The psychiatrist may ask whether the patient felt dazed, disoriented, or confused and order a CT scan of the patient's brain.

Finally, to see whether the patient's speaking behaviour is an instance of the psychiatric symptom of alogia, the following propositions were to be assessed:

- The patient shown relevantly worsened outcomes in a verbal fluency task.
- The patient shows no impairment in semantic memory.
- The patient shows no impairment in context processing.

To gather the information required to evaluate this set of diagnostic propositions and add it to the prepared description of the patient, the psychiatrist would have to conduct cognitive testing assessing the patient's verbal fluency, semantic memory, and context processing to assess the presence or absence of the required patterns of impairments in these functions associated with an impaired execution function of controlled retrieval, since these are considered to co-occur with the complaint of reluctant speech when it is a case of alogia.

Just as I laid out for the in-depth evaluation of three potential diagnostic evaluations potentially applicable to the complaint of reluctant speech, the same could be done for the other symptoms we are familiar with from the last section: the complaint of reluctant speech, for example, or the complaint of interpersonal problems that was discussed in the context of dimensional symptom diagnostics. But I think the principle is clear and that going through this in detail would be a rather repetitive exercise. I will therefore simply outline in brief how this task would be approached for these two examples.

To evaluate the complaint of constant worry again, the sets of diagnostic propositions would be evaluated by questioning the patient. In particular, questions might address his current life situation, to evaluate whether the patient's reactions are best understood as a non-pathological reaction to hostile living circumstances. Inquiries might also target the patient's worry-related experiences, such as the content of their experience (i.e., the content of cognitive states considered to be worry), their attitudes towards (or appraisal of) these experiences, and the patterns of occurrence of these experiences (i.e., under what circumstances, how often, and for how long they occur). The answers to these questions would in turn be relevant to evaluating whether the patient's complaints fit the criteria for compulsive thoughts or generalised anxiety. Rather than going into more detail on the evaluation of this and other complaints, I will now proceed to discussion of another topic: dimensional diagnosis and how it may be evaluated.

To evaluate the complaint of recurring interpersonal problems, regarding the possibility of whether it may be a case of personality pathology such as "moderate interpersonal problems", the psychiatrist would interview the patient and maybe people from the patient's social circle, if possible, to assess where their problems occur (in personal relationships and/or professional contexts), whether the patient is able to maintain friendships and if so how many, whether conflicts occur persistently or only occasionally or in specific circumstances, and whether the patient's romantic relationships seem to be constantly characterised by serious disruptions or are apparently showing the patient to be extremely submissive towards their partner. To gather the information that might be relevant to distinguish among alternative explanations, the psychiatrist would then also consider evaluating the patient's physical health by asking questions and potentially consulting medical records, to see whether the patient may suffer from any impairments that might impact social interaction. If any such impairments were found, the psychiatrist would inquire into how, if at all, the patient communicates with others about these impairments, as well as whether people who tend to have conflicts with the patient indicate that their problem with the patient results from an impression that might result from how the patient's impairments impact their communication behaviour.

Once diagnostic information has been collected, guided by the sets of propositions that are assumed to constitute the space of plausible evaluations of the presented complaints, the psychiatrist arrives at a prepared description of the patient. This description is prepared in that it resulted from the psychiatrist's skilful use of diagnostic tools to assess propositions about the patient, such that the resulting overall grasp of the patient's situation can be thought of as itself presenting a list of propositions stating facts about the patient, geared towards the purpose of comparing the diagnostic sets of propositions against the description of the patient to judge which of the patient's complaint, should, in accordance with the matching sets of diagnostic propositions, be judged to qualify as a psychiatric symptom, a psycho-behavioural aspect of a non-psychiatric disease, or just a distressing or unusual but not (psycho)pathological condition. However, the process of comparing the different sets of propositions to the prepared description of the patient's presentation is a relatively complex task; we need to look at it in some detail to do it justice.

By comparing different sets of diagnostic propositions to the patient's data to select which diagnostic evaluation a patient's complaint should receive, the diagnostic procedure in psychiatry, as well as in many other branches of medicine, regularly takes two forms: differential diagnostics and exclusion diagnostics (which is a special instance of differential diagnostics rather than a whole different type). Differential diagnostics as well as exclusion diagnostics are approaches to using sets of diagnostic propositions to assess initial complaints of patients in order to decide between different diagnostic interpretations of these complaints. Let me briefly present the basic idea behind these two approaches in diagnostics and then discuss another clinical example.

In differential diagnostics decisions between different diagnostic interpretations occurs by deciding which set of diagnostic propositions is best realised by the patient in the context of their complaint. The potential candidates will consist of only those diagnostic options whose diagnostic propositions suit the patient's presentation well enough to be appear plausible. From among these sufficiently well-fitting sets of propositions, the best-fitting one – that is, the one best supported by diagnostic information – will be selected to provide the diagnostic conclusion (i.e., the complaint X is the symptom Y or Z). When precisely a set of propositions can be considered to fit the presentation of a patient sufficiently well to be considered in principle applicable (if no other set of propositions fits the patient's presentation better) is a decision that must be made by the clinician in light of the assumed fidelity criteria for the specific diagnostic option, which (as discussed in 2.1) derive from the overall psychopathological background knowledge about the condition from which the set of propositions is derived.

Exclusion diagnostics takes place in a similar manner, with one exception. Exclusion diagnostics also compares sets of diagnostic propositions in a diagnostic evaluation of complaints against the patient's presentation. However, by contrast with an instance of differential diagnostics, an outcome that can also enable a diagnostic conclusion here is that none of the diagnostic set of propositions applies that would render the patient's presentation a certain type of psychiatric symptom, a psychological aspect of a non-psychiatric medical problem, or a non-pathological psychological distress. If this is the case, the complaint will usually be judged to be a psychiatric complaint, though one that lacks any deeper constitutive understanding.

In the case of exclusion diagnostics, the diagnostic label is chosen not because the patient's presentation matches up with a constitutive understanding of the corresponding psychiatric symptom, but also not because it was present as a complaint. In other words, the label is not positively identified due to its matching with a psychopathologically constitutive understanding of this psychopathological condition as opposed to some other pathological or non-pathological condition, nor is the evaluation provided simply because there was a report of a complaint. There is no direct inference from the presence of a complaint to the symptom diagnosis, but there is still relevant diagnostic effort in comparing alternative sets of diagnostic propositions to account for the complaint in play, although they all fail in the case of an exclusion diagnosis. Why is this important? Because this supports my point that diagnostic evaluations do not result straightforwardly from merely classifying complaints, but that there is always a layer of in-depth evaluation at work. Exclusion diagnostics occurs if no other set of diagnostic propositions appears to map sufficiently well onto the patient's diagnostic information to support a diagnostic inference that the psychiatric symptom in question is present, so that the absence of evidence for an alternative diagnosis is taken to support the classification as the remaining option for providing a diagnostic label.

Although the logic behind both inferential pathways is straightforward, another example for this instance of diagnostic practice might be useful. An example of exclusion diagnostics may be especially helpful, since imagining a differential diagnostic process based on my previous discussion should be straightforward. In the case of constant worry, or reluctant speech, I have rather extensively discussed their potential diagnostic evaluations and the sets of propositions that would be compared to the prepared description of the patient in order to decide which evaluation to choose. We can readily imagine how a comparative judgement of differential diagnostics would proceed: evaluating the applicability of all these sets of propositions, judging which of them in principle apply sufficiently well to embrace them, and then picking the one that best suits the patient's presentation. Exclusion diagnostics, on the other hand, seems to be a format of diagnostic inference that is less well covered by the basic setup of evaluating complaints that I provided in my discussion of screening, and it might therefore be harder to grasp.

To explore a case of exclusion diagnostics, let us consider the patient presenting with the complaint of hearing voices in the absence of someone speaking. In recognition of this complaint, the psychiatrist will consider different potential diagnostic evaluations. On the one hand, hearing voices may be an instance of auditory hallucinations as a psychiatric symptom that occurs, for example, in the context of schizophrenia, bipolar disorder, schizoaffective disorder, severe depression, borderline personality disorder, and post-traumatic stress disorder (Choong, Hunter, and Woodruff, 2007; Waters et al., 2017). Alternatively, we know that hearing voices also occurs outside the realm of psychiatry narrowly understood – that is, in cases that would not count it towards being a symptom of (for example) schizophrenia. Hearing voices can be the result of (among others) acute sleep deprivation (Waters et al., 2018), malnutrition (Mittal, 2010), coeliac disease (Lindberg, Marco, and Klas, 2013), brain tumours (Madhusoodanan et al., 2004), certain forms of encephalitis (Silva and Brucki, 2010; Boyd et al., 2013; Kayser et al., 2013), traumatic brain injury (Sachdev, Smith, and Cathcart, 2001), sensory deprivation (Mason and Brady, 2009); it can also be a side-effect of prescription medications (Abou et al., 2015) or substance abuse (Fiorentini et al., 2021). The complaint of hearing voices might also occur as a normal phenomenon, such as in hypnagogic and hypnopompic experiences (Ohayon et al., 1996; Waters et al. 2016).

We have some understanding for such non-psychopathological circumstances that may accompany the complaint of hearing voices. In terms of the here discussed approach to diagnostics that means we can draw on sets of constitutive propositions that would enable us to evaluate them as occurring in the context of the system's execution of the disposition to hear voices. In context of such evaluation it would be judged that this disposition is actualised so that it would not be psychopathological symptom. For example, we know that hearing voices as a hypnagogic/hypnopompic experience occurs only during the transitions from wakefulness to sleep or the other way around, respectively; that people usually know that they are not real; and that the experiences are fleeting and are interrupted if one is fully woken. Things look different in our current understanding of auditory hallucinations as sign of psychiatric disorders. Outside of the contexts that would lead to its evaluation as a non-psychiatric medical complaint, a non-pathological sign of psychological distress, or just a normal psychological occurrence, voice-hearing is poorly understood. Commonsensical positive characterisations of voice-hearing – beyond the complaint-level description that characterises psychopathological instances of it – are hard to come by. Under circumstances in which the psychiatrist does not have a way to evaluate each potential diagnostic option *qua* testing a set of propositions that should state facts about the individual suspected to suffer from a certain condition, the psychiatrist instead evaluates each diagnostic option that he has sets of propositions for. If none of these applies sufficiently well to the patient's case to provide a potential basis for drawing a positive diagnostic inference to the applicable evaluation, the remaining option is a psychiatric symptom (i.e., *auditory verbal hallucinations*), which provides the result of the diagnostic evaluation of the complaint.

Having considered the step of in-depth evaluation in general and discussed its occurrence in differential and exclusion diagnostics, I now want to come to the second task of this section: showing that this step of diagnostics allows me to map criterion (*iii*) (model/world comparison) and criterion (*vii*) (error-model based differential and exclusion diagnostics) onto this step of the process. The success of this step is crucial, since it will complete the list of criteria given in the Introduction and therefore show that this part of diagnostics can be understood as qualitative, constitutive diagnostic modelling.

3.2.2 In-Depth Evaluation as Modelling

It is straightforward to show that diagnostic in-depth evaluation can be considered model/world comparison (criterion (*iii*)) and error-model based differential and exclusion diagnostics (criterion (*vii*)). If we consider the theoretical structures used by clinicians in the in-depth evaluation – that is, the sets of diagnostic propositions – what they do with these structures is compare them with relevant information about the system that provides insight into the actual the patient regarding those aspects that are targeted by the theoretical structure. By collecting this information and putting it together in an overall description of the patient containing the information relevant to assessing the relevant diagnostic models against the patient, clinicians compare the propositional structures they construed and analysed in the context of the screening phase with the real-world system they were set up to target in the context of the diagnostic process. By so doing, they can judge which of the propositional theoretical structures (if any) match with the real-world system and decide based on the outcome of this process which models to accept and which models to reject based on established fidelity criteria. Thus, after what I argued should be understood as the steps of model construal and analysis, psychiatrists now carry out the matching process described in the last chapter as the act of model/world comparison. There, model/world comparison was introduced as precisely this: the last step of the modelling process in which a theoretical structure – one that is intended to target a real-world system and that was previously set up in a procedure of model construal and analysis – is compared to the real-world system considering the intended assignments, fidelity, and scope of the theoretical structure, to decide whether the structure maps well enough onto the chosen real-world target to be considered permissible. Given this characterisation, and assuming the success of my earlier attempts to demonstrate that a process that can be thought of as model construal and analysis takes place previously in the diagnostic process, I consider it fair to conclude that in depth-evaluation can be considered to exemplify model/world comparison. Thus criterion (iii) is fulfilled, which was the last criterion needed to complete the selection of criteria (i), (*ii*), and (*iii*) needed to show that there is a modelling process taking place in the diagnostic reasoning process. Next let me turn to criterion (*vii*).

The diagnostic reasoning process taking place in psychiatric diagnostics meets criterion (*vii*) and thus uses diagnostic sets of propositions (considered as models) in a diagnostic modelling process that employs differential diagnostic and exclusion diagnostic modelling. As described in the last subsection, the in-depth evaluation tackles the diagnostic evaluation of complaints recognised in the screening phase with the help of sets of diagnostic propositions based on previously recognised complaints. To this end, the clinician collects information about the patient suitable for evaluating these propositions in order to generate a prepared description of the patient. Once all information is there, the psychiatrist begins to compare the propositional diagnostic models against the prepared description. As I described, this happens in two *modi operandi*, differential diagnostics and exclusion diagnostics, where exclusion diagnostics is one path that differential diagnostic may turn out to take, rather than an independent approach.

Considering what has been said about model-based diagnostics, more particular model based differential and exclusion diagnostics, it seems that the way in which propositional models are illustrated to be used by psychiatrist for psychiatric differential and exclusion diagnostics match up neatly. If we think of the complaint recognised in the screening process as constituting the *prima facie* error, which is the starting point of the diagnostic modelling process described in the last chapter, and of the sets of diagnostic propositions used by psychiatrists as diagnostic models, then the diagnostic process of differential diagnostics in psychiatry – just like the differential diagnostic modelling procedure – is meant to be a comparison between different models in light of a description of the target system. The models are meant to match in order to allow the *prima facie* error produced by the system to be classified as a certain type of error. This match also seems to occur for exclusion diagnostics. In exclusion diagnostics, the sets of propositions compared to relevant features of the patient not leading to a match between one of the sets of propositions enables the exclusion diagnostic determination of the classification of the patient's complaint. This matches up with what was set out in the last chapter about exclusion diagnostic modelling. There, I discussed how in exclusion diagnostic modelling, after the recognition of *prima facie* errors, the modeller will compare a selection of diagnostic models against the system producing the *prima facie* error, reserving a specific label in the diagnostic taxonomy to be applied to the system in case none of the diagnostic models matches with the system. In conclusion, these parallels seem to warrant the conclusion that the in-depth evaluation process meets criterion (*vii*) and thus embodies the use of differential and exclusion diagnostic modelling.[14] This means that all criteria I set out in the Introduction are now mapped onto the steps of the diagnostic psychiatric process that have been discussed so far. We can therefore understand psychiatric diagnostics as following the method of qualitative, constitutive diagnostic modelling as described in the previous chapter. However, the parts of diagnostic process that have been considered so far are not the whole story.

The remaining aspect of the diagnostic process and the reasoning process that psychiatrists go through to complete it has not been covered by the modelling procedure outlined so far. However, the rest of the process, namely the final step of for-

14 We covered the instantiation of both pathways of error-model based diagnostic decision processes at the end of a psychiatric evaluation or in model/world comparison. The remaining form of diagnostic conclusion-drawing that I discussed in the last chapter but not here is *normative-model based diagnostic conclusion-drawing*. An instance of this kind of diagnostics would involve the psychiatrist recognising a patient's complaint and making a straightforward inference from the recognised complaint to the diagnostic judgement of the presence of a psychiatric symptom, since the initial complaint occurs so unambiguously only in the presence of this symptom that the complaint automatically has a symptom value – i.e., the complaint is the symptom. In terms of modelling, this would mean that based on the normative model, an initial error in the system is recognised that is so unique that it only allows for one diagnostic evaluation, making further investigation (i.e., in terms of error models) unnecessary. To my knowledge and judging from discussions with other expert clinicians, there are no cases like this in diagnostic evaluation if it is carried out properly and with expert clinical knowledge. Counterexamples I have been presented with so far seem to suffer from the problem that they consider as part of the initial error recognition or the complaint registration information that according to my description would be part of the in-depth evaluation. Accepting such cases would collapse steps that I am trying to keep distinct in my approach, and would confuse normative-model based diagnostics with error-model based diagnostics and initial recognition of errors with deep analysis of a system.

mulating the diagnostic proposal based as an outcome of the in-depth evaluation, capitalises heavily on the process already described. This is why I call my overall proposal the model-based approach. In the next section, I will look at the last part of diagnostic conclusion-drawing and propose how to understand this part in terms of the ways in which it draws on the previous modelling process and also goes beyond it to propose a final syndromal diagnosis in accordance with diagnostic manuals.

3.3 The Diagnostic Proposal as a Synthesis of Modelling Outcomes and Pattern Recognition

As the psychiatrist ends their diagnostic information-gathering and makes up their mind about which of the patient's complaints should be evaluated as which kind of psychiatric symptom (or alternatively as a psycho-behavioural problem associated with a non-psychiatric medical problem, or a distressing but not pathological psycho-behavioural complaint), it is time to provide a diagnostic proposal. The diagnostic proposal, as discussed in the first chapter, contains two elements: the case formulation and the syndromal diagnosis.

In the case formulation, the psychiatrist organises the diagnostic information that has been obtained and the diagnostic evaluations that are supported by them in a way that allows the reader to understand the evidence determining which of the patient's complaints were given which diagnostic evaluation. The mental construction of the case formulation provides and makes transparent the justification for the symptom-related decision, and therefore also indirectly the justification for the disorder-diagnostic decision that must then be made. As such, the case formulation will contain information regarding all differential or exclusion diagnostic decisions and will therefore present an overall propositional representation of the psychiatrist's diagnostic understanding of the relevant findings in the patient's case, such that it becomes clear which findings about the patient led to the evaluation of which complaint as which symptom (or non-symptom). The written version of the case formulation that may become part of the patient's medical file, or that the psychiatrist may present at a case conferences among colleagues, will often be shorter than this full-fledged version for pragmatic reasons such as limits on time and space. This version will, for example, often only contain information about which propositions were found to apply and contribute to the resulting evaluation of each complaint, and no information about precisely why every other diagnostic possibility that was taken into consideration was ultimately rejected.

As part of the case formulation, we also sometimes see working hypotheses about causal connections between conditions – that is, relationships to be proposed between the patient's symptoms. In cases where a proposition supporting a diagnostic evaluation appears to be causally related to the occurrence of other propositions, this hypothesised but not evaluated causal relationship may be included as a hypothesis in the formulation. For example, a patient may suffer from sleeping problems (initially considered as potentially the psychiatric symptom of *insomnia)* that are evaluated not to be a psychiatric symptom, since it was decided through the in-depth evaluation that no psychiatric problem are present, but another sleep-disrupting medical problem (e.g., chronic pain[15]) is present. Likewise, the patient's loss of interest and pleasure in activities he once liked (initially considered as potentially indicating *anhedonia*) is evaluated as resulting from a lack of energy due to lack of sleep. In this case, the psychiatrist might point out the assumed causal relationship (chronic pain sleeping problems low energy lack of interest) in the case formulation.[16] However, psychiatrists will not find more or less causal proposals for all relations amongst symptoms; therefore, this aspect of the case formulation is contingent. Thus, on the level of the case formulation, the symptoms or otherwise classified complaints assumed to be present in the patient are also related causally in ways that occur plausible as part of the process of synthesising information from diagnostic evaluations.

The case formulation developed in this way does, as required by the DSM (see the discussion in Chapter 1), explain the patient's psychopathological status in two ways. On the one hand, it offers the more robust, vertical method *qua* constitutive explanation. This is based on the evaluation of constitutive models of the conditions judged to be present against the patient's presentation. The results of this modelling feed up from the lower-level diagnostic modelling process into the case formulation (hence is vertical). On the other hand, the case formulation offers the weaker, more speculative, horizontal causal consideration, which may be proposed by the psychiatrist without specific evaluations against the actual presentation of the patient, but solely by recognising the present complaints and perhaps by drawing on some information about the order in which they occurred (since temporal order allows for a plausible suggestion that one problem might be the cause of another) and the time at which any of the complaints got better or worse (since associations between improvement and worsening of complaints could indicate causal connections). Thus,

15 This is a problem reported by more than 50% of patients suffering from long-standing sleeping problems (Ohayon, 2002, 2005).

16 Pointing out such suspected causal relationships is reminiscent of an approach to psychopathology that has been growing in popularity: the network theory of mental disorder (e.g., Borsboom, 2017). Roughly speaking, this theory attempts to understand mental disorders as networks of symptoms causing and perpetuating each other, and offers clinicians ways to intervene in disorders by addressing specific causally relevant nodes in the network. While Borsboom and others have made great efforts to develop this approach into an empirical research paradigm, I think it is fair to say that clinicians have been thinking in this way in the context of case formulations and their uses throughout the entire history of literature on diagnostic case formulations.

relationships on the (horizontal) level of complaints that seem plausible can be articulated. The case formulation therefore explains the present condition primarily *qua* the constitutive explanatory support that justifies claims about which conditions is claimed to be present in the patient. Secondarily the case formulation provides speculative hinges on the causal relationships between aspects of the presentation. In so doing, it does what the APA (2013) requires. It "recognize[s] when the combination of predisposing, precipitating, perpetuating and protective factors […] [have] resulted in a psychopathological condition" (ibid., p. 19) by making use of such factors in the context of the constitutive models used to identify psychiatric symptoms, and by also allowing for some causal speculations regarding "social, psychological, and biological factors that may have constituted to developing a given mental disorder" (ibid.).

With the case formulation as the summary of the overall diagnostic evaluation via diagnostic modelling that specifies the psychiatric symptoms of the patient, plus some causal speculations expressed in it, the psychiatrist can also provide the syndromal diagnosis. Based on the list of identified psychiatric symptoms backed up by the case formulation, the psychiatrist will select a syndromal psychiatric diagnosis according to the rules of the diagnostic manual in use, currently usually either the DSM-5 or the ICD-10 or ICD-11. While someone new to psychiatric diagnostics will not be familiar with the exact criteria of each diagnosis, the trained expert familiar with the manual will be able to make this inference from the list of recognised psychiatric symptoms to the correct manual-based diagnosis relatively effortlessly. The patient may thereby receive one diagnosis or – not uncommonly – multiple diagnoses.

The syndromal diagnosis whose attributions are justified by the presence of clusters of symptoms, which in turn are justified by the model-based attribution of symptoms, then relates to the modelling process, which provides the ultimate justification. The attribution itself, however, is itself not modelling but a rather straightforward recognition of certain patterns of required symptoms plus the consideration of certain additional rules for diagnostic attribution – for example, that some diagnoses will not be given if the criteria for others are fulfilled. If someone fulfils one of the potential sets of criteria for a major depression, they will also meet all criteria of a mild depression, but in this case one is supposed to diagnose only the major depression. Or if criteria for an anxiety diagnosis and a depression diagnosis are both met, one is supposed to minimise the diagnostic entities attributed and instead of diagnosing both, one should rather opt for a diagnosis of depression with anxiety features. However, despite these additional rules on the level of syndrome attribution, which are meant to foster parsimonious attribution of syndromal entities, the process is otherwise straightforward for anyone who has learned to match symptom patterns with syndromes by heart. To capture this aspect of the diagnostic process, which is based on the outcomes of the diagnostic machinery of modelling that is feeding in information about recognised symptoms, I propose – in line with research on medical cognition (e.g., Conderre et al., 2003; Groves, O'Rourke, and Alexander, 2003; Loveday et al., 2013) – to understand it as a form of prototype-based pattern recognition. What do I mean by this?

The prototype theory of pattern recognition in cognitive psychology is a model of pattern recognition as a cognitive process, according to which different prototypes of objects are memorised by the system:

> in the process of pattern recognition, outside simulation only needs to be compared with the prototype, and the sense to objects comes from the matching between input information and prototype. Once outside simulating information matches best with a certain prototype in the brain, the information can be ranged in the category of that prototype and recognized. (Pi et al., 2008, p. 435)

If we are understanding the process by which, from a list of symptoms, psychiatrists infer matching syndromes, this would mean that they know the relevant combinations of symptoms that would support a disorder diagnosis as prototypical patterns of symptoms that are inferred once the outside information (i.e., a recognised pattern of symptoms as proposed by the case formulation) is recognised. This recognition then cues the relevant prototype that is associated with the diagnosis matching the pattern of symptoms.[17]

As a result, the process of setting up the diagnostic proposal consists in the cognitive synthesis of the information that is generated as an outcome of the in-depth evaluation, especially the information in which complaints of the patient are evaluated as psychiatric symptoms. In addition to this synthesis, which presents the selected models for the patient's complaints, the psychiatrist may offer potential causal interpretations of relationships between different facts about the patient that were found to hold true in the evaluation of diagnostically relevant propositions, so as to add an extra, though usually rather speculative, layer of causal explanation to the diagnostic outcomes whose primary justification is constitutive. Subsequently, the list of attributed symptoms is used as a baseline of questions that – for psychiatrists who are well trained and aware of the pattern's symptoms and the constraints of potential additional requirements and diagnostic rules (e.g., mutually exclusive

17 If knowledge about prototypes is lacking in a psychiatrist – for example, because they are a novice or there have been recent changes in diagnostic manuals (or because they just want to double-check the criteria) – the automatic cue-based procedure may also be turned into a lookup-and-match process in which the psychiatrist recognises the pattern in a more effortful comparison between a certain set of symptoms and the diagnostic manual's requirements for symptom combinations sufficient for a diagnosis. However, in this context I assume that diagnostic experts are aware of diagnostic rules and of sufficient combinations of symptoms for a given diagnosis.

diagnoses, diagnostic label minimisation) – can then be used to rapidly recognise the potential prototypically occurring patterns of symptoms that support one syndromal diagnosis over another and infer its presence according to the diagnostic manual. Thus, the diagnostic case formulation as well as the disorder diagnostic proposal are produced as the outputs of the diagnostic procedure.

3.4 Conclusion

This chapter has synthesised the material from the previous two chapters on psychiatric diagnostics and diagnostic modelling, arguing that the process of psychiatric diagnostics can be largely understood as a qualitative, constitutive modelling process followed by an inferential procedure that relies on pattern recognition. To support this argument, I proposed seven criteria that map onto the diagnostic process. I then provided a detailed discussion of the diagnostic screening procedure and the in-depth evaluation, using examples to illustrate how these steps and the diagnostic reasoning guiding the transitions between them exemplify the proposed criteria. In the final section, I discussed the diagnostic proposal, which consists of the case formulation and the syndromal disorder diagnosis. I explained how the case formulation is derived from previous modelling efforts and potentially influenced by causal considerations. I also discussed how the inferences from recognised psychiatric symptoms are made using a straightforward pattern recognition procedure in accordance with the diagnostic manual's rules, which dictate the prototype knowledge presentations regarding symptom patterns that psychiatrists keep in mind when determining which syndromes to diagnose based on the list of present symptoms.

This chapter marks a significant step in my attempt to present and defend my answer to the Methodological Question. By showing how my primarily model-based proposal meets the first requirement of the Methodological Question and provides a descriptively adequate account of the basic process of psychiatric diagnostic reasoning, I have set the stage for fulfilling the second and third requirements. These tasks were already addressed in the previous chapter, where I discussed the inferential strategy of model-based diagnostics and its justification. Given that the overall description of diagnostic reasoning presented in the previous chapter applies to psychiatric diagnostics, it follows that the other aspects of the methodology also apply. Therefore, this chapter establishes the applicability of the proposed methodology of qualitative, constitutive diagnostic modelling to psychiatric diagnostics, and also, with the pattern recognition proposal, offers a straightforward account that complements the modelling procedure to explain the rest of the diagnostic process, resulting in a compete answer to the Methodological Question.

Having presented my proposal for answering the Methodological Question, which includes a description of the process, a rationale for the inferential procedure, and a discussion of its justification, I now aim to demonstrate that my proposal goes beyond mere adequacy and satisfies the additional criteria for a good answer to the Methodological Question outlined in the Introduction to this thesis.

5. Fulfilling Desiderata

In the preceding chapters, I developed my model-based account of diagnostic reasoning in psychiatry. In this chapter I want to let it do some work by showing that it not only meets the adequacy conditions to for an answer to the Methodological Question, as suggested by the end of the last chapter, but in addition fulfils the desiderata that I set out in the Introduction. These desiderata were that the proposed answer to the Methodological Question should:

1. provide a comprehensive account of the core aspect of the process of psychiatric diagnostic reasoning
2. present a proposal to us that is cognitively realistic, thus can take place in actual diagnostic efforts
3. make sense of the difference between misdiagnosis and diagnostic malpractice in psychiatry
4. explain the occurrence and resolution of diagnostic uncertainty in psychiatric clinical diagnostics
5. explain the phenomenon of good instinctual diagnosis and what is problematic about it
6. explain the occurrence and resolution of diagnostic disagreements over time within and between experts
7. provide guidance for thinking about how changes in psychopathology may be integrated with or change the methods of diagnostic reasoning.

These desiderata were proposed to be relevant to address in a proposal for answering the Methodological Question since they show that the proposal is a helpful guide either to attaining a basic grasp of psychiatric diagnostics itself, or to understanding more specific aspects of (and phenomena in the context of) diagnostic reasoning that are commonly encountered and thus useful to explain. Let us briefly recap the relevance of each of the desiderata.

A proposal for answering the Methodological Question should ideally provide a comprehensive account encompassing all aspects of the diagnostic process and leaving no central aspect unexplained. It should ensure that its proposal is within the general capacities of a psychiatrist to be carried out as a realistic person-level cognitive process, and thus can be taken as a realistic method (i.e., a learned belief-forming procedure) that psychiatrists may pursue in their everyday diagnostic clinical work. The proposal should also enable us to understand the occurrence and resolution of diagnostic uncertainties and disagreements. Mistakes in diagnostics unfortunately occur and differentiating between mere misdiagnosis and actual malpractice is of high ethical and legal relevance. To get a hold on "diagnostic instincts" seems important since everyone who has ever worked in a clinical context will have seen experienced clinicians shooting diagnostic guesses from the hip who, more often than not, seem to be right, so that it is relevant to have a well-founded attitude towards how this form of diagnostics works and why it is (or is not) credible. Finally, to make sense of the possibility of integrating into diagnostic practice ongoing changes in our understanding of psychopathology, as well as to speculate as to what the future of diagnostics might mean for our current methos of diagnostic reasoning, is central to showing the theory's plausibility in terms of its responsiveness to change. It should be robust in that it allows us to explain how current diagnostic reasoning integrates minor changes, but sensitive enough to large-scale changes to diagnostics to be falsifiable, otherwise it would be too generic. In the following section, I will discuss how my answer to the Methodological Question enables us to meet all the desiderata listed above.

4.1 Comprehensiveness

For a proposal to address the Methodological Question in a comprehensive manner requires it two do two things. It requires the proposals descriptive suggestion of a method as part of the methodology to leave no relevant aspect of the diagnostic reasoning process unaddressed and to make sense of its different aspects with a reasonable degree of detail. To meet these two requirements is what would make the proposal comprehensive. Whether my own proposal, the model-based account of psychiatric diagnostic reasoning, meets the criterion of comprehensiveness depends on two things. First, it depends on whether one accepts my basic account of the process of clinical psychiatric diagnostics as the proper core procedure of contemporary diagnostic reasoning, as presented in the first chapter and via a more example-oriented treatment in the third chapter. Second, meeting this criterion depends on whether one accepts that the attempt to map my understanding of diagnostic modelling as laid out in the second chapter, plus my limited additional remarks about how the case formulation (as a composition of modelling outcomes) and the disorder diagnosis (as pattern recognition) maps onto the described process of clinical psychiatric diagnostics indeed explains the described diagnostic reasoning process on a sufficient level of detail. The reasons why I believe that my presentation of the clinical diagnostic process is adequate were presented in the first chapter, and the considerations that make me think that the proposed method of modelling and pattern recognition maps onto psychiatric diagnostics, have just been laid out in Chapter 3, so I will simply reiterate my previous points here in a more abstract fashion.

The first aspect of ensuring that my proposal to address the Methodological Question meets the criterion of comprehensiveness involves checking that I provided my attempt to answer it with an adequate starting description of psychiatric diagnostics – in other words, a description that itself has an adequate scope and explores the process in relevant depth. To ensure that it has an adequate scope, as discussed in more detail in earlier chapters, I considered a recent edition of widely regarded psychiatric training literature that is intended to lay out the general core procedures of clinical psychiatric diagnostics, as well as recent guidelines of psychiatric expert societies. Focusing on those sources was meant to ensure the proper scope for what I consider to be the constitutive core procedures of proper, contemporary, clinical psychiatric diagnostics. While my approach to account for the overall diagnostic procedure in Chapter 1 did not delve into too much detail for specific cases but rather provided an overview, Chapter 3 provided several clinical examples in line with my general understanding in a more illustrative fashion. This more detailed presentation in Chapter 3, with the more general architecture from Chapter 1 in the background, provided a foundation on which I then attempted to demonstrate the mapping between my model-based proposal and pattern recognition in the diagnostic process.

To ensure that my efforts to establish my proposal turn out to be a comprehensible account of psychiatric diagnostics, I went through all phases of the diagnostic process initially identified in Chapter 1 to map onto it all aspects of the method I had claimed take place. Thereby I outlined how we should understand the relevant facets of each stage of psychiatric diagnostics in light of the method I proposed. To briefly take one example, I opened my discussion with the first step of the diagnostic process, the screening phase. This phase is meant to enable the psychiatrist to recognise a patient's complaints based on previous assumptions about what is within the range of normal psycho-behavioural features, such that deviations of a patient from these assumed states might indicate the presence of a psychiatric symptom and are thus identified as complaints, which further down the road, in the in-depth evaluation, will be evaluated to decide whether they are indeed a psychiatric symptom or not. I illustrated this step in detail, moving from a generalising description of this step to concrete clinical examples. In my attempt to map diagnostic modelling onto psychiatric diagnostics, I proposed that this step in the diagnostic process and its different aspects is equivalent to the initial error-recognition step in diagnostic modelling. I argued that the background assumption of the psychiatrist to discover complaints equals the normative model based on which initial error recognition identifies *prima facie* errors, and that the complaints identified by the psychiatrist equal these *prima facie* errors, being discovered using the normative model and later evaluated via the diagnostic procedures. I then discussed the realisation of this process in the concrete clinical examples I provided.

This exemplary step from my work shows that I described the clinical diagnostic process in general terms, to a degree of detail where the next best step to offer further detail was to provide concrete case examples. In other words, I described the clinical diagnostic process to the lowest still general level of detail in which I could describe it before transitioning to single cases. It therefore seems that the mapping of the method onto the process whose description is provided on this level of detail is as comprehensive as it can become before forfeiting its claim to allow us to discuss the diagnostic procedure in general. Hence the discussion of stages, aspects, and the functional connections between them in psychiatric diagnostics, and the fact that everything I claimed about diagnostics was mapped onto my application of the method of model-based diagnostics (just as I did in the brief excerpt of my efforts just discussed), together seem to justify the assessment of my answer to the Methodological Question as comprehensive.

4.2 Cognitive Realism

To ensure that an answer to the Methodological Question is not only in principle adequate to match the requirements to qualify as an answer to the Methodological Question, it should also be realistic – that is, be a procedure that could plausibly be carried out as a learned person-level procedure by real clinicians doing diagnostic work. Only then can it qualify as a method (i.e., a learned belief-forming procedure) that could be the actual cognitive work undertaken by clinicians. In other words, the proposal should be cognitively realistic.

To see whether my model-based account presents a realistic proposal, we need to ensure that it proposes a format of reasoning that seems to equal what commonsensically takes place in clinicians' minds when they think about their patients. Regarding the requirements on information-processing, the amount should not exceed what can plausibly be assumed to be within the capacity for cognitive load of diagnostic experts. In addition, since a method is (as discussed in the Introduction) a learned belief-forming procedure, it should be *prima facie* realistic that the way diagnostics take places according to the proposed method, and thus following the rules of the method, should be something that can plausibly be learned.

Let us begin with the format. The chief format of diagnostic reasoning that I am proposing is qualitative reasoning in the form of propositions that contain diagnostically relevant information. *Prima facie* this seems to fit well with what psychiatrists do. As I said earlier, clinicians do not calculate the diagnoses of their patients. Rather, when we look at conversations between clinicians speaking about patients, or when, as discussed earlier, we look at diagnostic exercises or research involving diagnosing clinicians using think-aloud protocols, we usually find them engaging in diagnostic reasoning in terms of normal-language sentences, describing diagnostic requirements as well as information about the patient and deciding which of these propositions apply and what to infer from that. It thus seems that my account considering propositional models as information bearers and as vehicles of diagnostic reasoning matches well with what we find in clinical diagnostics, when it comes to describing the process on the personal level of the psychology of clinicians and their intentional efforts to evaluate patients.

When we think about the cognitive load associated with this proposal, it seems bearable. Of course, the psychiatrist does not have all potentially relevant propositions that might become relevant in the diagnostic process present in their working memory at the same time, but they are present in the background knowledge base resulting from the psychiatrist's education. When carrying out the screening procedure, for example, psychiatrists systematically explore the different aspects of the patient's life, bearing in mind the propositions of the aspect of the normative model that is being compared with the patient's psycho-behavioural functioning in this area. If the patient spontaneously reports complaints, the psychiatrist entertains the normative propositions relevant for the relevant aspect of the psycho-behavioural presentation of the patient and compares the complaint with the propositions. The same goes for the diagnostic propositional models. The psychiatrist never has all of them at the forefront of their mind all the time, but a recognised complaint will trigger the recall of potential diagnostic options that are all connected with diagnostic model structures whose content can be entertained and used to guide in-depth evaluation if needed. Furthermore, the inferences from present patterns of symptoms to an adequate diagnosis are (if carried out by a clinician who has learned the diagnostic manual) made not by calling to mind all disorders and their symptoms, but by recalling the adequate disorder diagnosis based on a certain set of previously identified symptoms present. Thus, cognitive load is managed by bringing only what is needed into the psychiatrist's immediate cognitive workspace. This management process is further supported by documenting (taking clinical notes on) steps of the diagnostic process to ensure that once made, inferences and their outcomes do not get lost.

Finally, the overall intentional person-level procedure of diagnostics that is carried out in this way also appears to be something that can be learned and that thus qualifies as a method. Nobody is born a diagnostic expert. Psychiatrists acquire their psychopathological and general medical knowledge base through their studies and clinical experiences and learn how to use it in a diagnostic process by consulting training literature and gathering clinical practice in which they are supervised in carrying out the stepwise process. They are taught what information about the patient may indicate which psychiatric or medical problem, what further information is needed to assess these options, and how they can generate this information in contact with the patient. All this and the further steps of the overall diagnostic procedure are taught to psychiatrists, which is possible because they can be told what to consider and which actions to take and not take as part of the diagnostic process. Because they can express what they had in mind when they attempted to provide a diagnosis and what their reasoning was in considering option (a) rather than option (b), they can be corrected in their reasoning and action, and so come closer to embodying the proper method of psychiatric diagnostics. Although nobody tells psychiatrists about normative models, propositional diagnostic models, or *prima facie* errors as part of their education, and they thus do not learn the method on a theoretical level, they do learn to carry out the diagnostic procedures such that by following these procedures they indeed follow the standards of the method of proper clinical diagnostic reasoning.

In sum, it seems that my proposal of the model-based account manifests all aspects of cognitive realism. It requires a plausibly manageable format and cognitive load from clinicians, and it appears that the method used is something that can be learned as part of clinical training. Thus, the desideratum of cognitive realism is fulfilled.

4.3 Misdiagnosis and Diagnostic Malpractice

Medical diagnosis is fallible. A diagnosis given to a patient by a diagnostic expert in any field of medicine can be wrong. The reasons why a wrong diagnosis can be made are numerous, from accidental documentation mistakes to mixing up test results, and from lack of scrutiny in examining a radiographic assessment to a blood test that against all the odds repeatedly yields false negatives. Some reasons why diagnostics may fail (such as mixing up results) can occur across many fields of medicine, while others (such as the failure to spot something important in a radiographic assessment) are more specific to certain medical disciplines. But independent of the medical discipline we are looking at, we may initially distinguish two general types of wrong diagnosis. I will label the first type *misdiagnosis* and the second type *diagnostic malpractice*. If a wrong diagnosis is a result misdiagnosis, the diagnosis was provided in accordance with the standards of diagnostic procedures and reasoning but the resulting diagnostic conclusion eventually turns out to be false. A wrong diagnosis resulting from malpractice, on the other hand, is one that results from a procedure of diagnostic reasoning that was not pursued in accordance with the standards of diagnostic reasoning.[1]

1 There are some complexities related to the notions of misdiagnosis and diagnostic malpractice. Misdiagnosis seems to be conceptually more closely linked to wrong diagnosis than to

To keep these two sources of error conceptually distinct and to know how to identify them is important for normative reasons. If someone follows the correct diagnostic procedures providing the standards of good diagnostics, arriving at a wrong diagnose is upsetting, but intuitively it seems that such an outcome is not the personal fault of the diagnostic expert. Imagine that the gold standard for diagnosing depression were a saliva test with a 0.1% false positive and false negative rate. If the diagnostic expert uses the test correctly, and the result is positive although the patient (as it turns out later) is not depressed, it seems that this is not the fault of the expert (who did as well as he could), but a risk inherent to the testing procedure. In cases of this kind, the diagnosing clinician would not be at fault or responsible for the wrong diagnosis or its immediate consequences. If, on the other hand, the wrong diagnostic result is attributable to mistakes made by the diagnostic expert in the diagnostic process that is under their control, things look different. In such a case, the clinician would arguably be at fault and responsible because they could have prevented the wrong diagnosis by following the standards of their profession.

Beyond just knowing who to blame, being able to differentiate between malpractice and misdiagnosis is important for legal reasons because malpractice, in contrast to misdiagnosis, is a legally relevant error that might grant patients the right to receive financial compensation and might cost a malpractising clinician their licence. Identifying such cases is also important for generating statistics on where and how often malpractice occurs, as well as for assessing the need for educational or administrative programs to prevent malpractice.[2]

malpractice. If someone is misdiagnoses, the diagnosis will necessarily be false. If someone receives a diagnosis via malpractice, this diagnosis might nonetheless be right by accident. However, even if a malpractising clinician is lucky and provides the right diagnosis, this would be considered problematic because they are not practising according to medical standards, which – independent of the outcome of their practice – is an issue, since there is an agreement to practise according to such standards in order to ensure quality care. So even if malpractice leads to the right result, there is reason to criticise the malpractising clinician. In the following, I will focus on malpractice with a wrong diagnostic outcome since these are the instances in which identifying and differentiating between malpractice and misdiagnosis will be of most relevance, at least legally, due to the (potential) cause of harm.

2 This understanding of malpractice is generally in line with the way it is treated in common law jurisdictions. Although details of the law differ significantly between different countries, in general, liability for malpractice in medical professions is given if there is a failure to show a fair, reasonable, and competent degree of skill, measured by the standards of the profession, and/or there is a violation of ethical standards (Giesen, 1988). A difference between most understandings of malpractice in law and my understanding is that there is often an additional *harm condition*. Only if the behaviour of the clinician caused significant harm to the patient will it qualify as malpractice. Although this may be a reasonable approach for the purpose of lawsuits for practical reasons (e.g., saving court resources, determining compensation), I think it is unreasonable to accept this consequentialist condition when we are discussing the nature of malpractice. The fact that the clinician enjoyed the moral luck that their be-

A theory of psychiatric diagnostic reasoning should provide the resources to make sense of this distinction between malpractice and misdiagnosis and provide guidance on how to identify malpractice in the context of psychiatric diagnostics. In the following, I will discuss how the model-based account does this. Let us start with misdiagnosing.

In short, misdiagnosis happens if the clinician follows best practice of diagnostic reasoning and nonetheless ends up providing a wrong diagnosis. How may misdiagnosis occur, according to the model-based account? Let us look at the diagnostic process as understood in the model-based account to try to spot the places where error leading to wrong diagnosis may occur, even if good practice has been conscientiously pursued. As we may recall from previous chapters, to carry out a proper diagnostic procedure the psychiatrist will have listened to the spontaneous complaints of the patient and systematically evaluated their psychopathological status. After so doing, the psychiatrist will have considered the different potential models of psychopathological, other medical, or non-medical conditions the patient may present accounting for their complaints. Then, by interviewing, testing, and examining the patient, they will gather the information that is relevant to evaluating the models of these conditions against the patient's presentation. Once the information has been gathered, the best-fitting (and sufficiently well-fitting) models for the present complaints will be selected, one or more diagnoses will be attributed to the patient based on the classification rules of the manual being used, and a case formulation will be provided. Assuming that all these steps are carried out adequately by the psychiatrist, there are two remaining loopholes that may promote wrong diagnosis. Both relate to the problem of insufficient information as the basis of the diagnostic reasoning procedure.

The first reason for misdiagnosis is *diagnostic uncertainty* resulting from ambivalence between multiple diagnostic options, because the information is insufficient to make a clear decision, potentially leading to a wrong diagnostic conclusion. As the topic of diagnostic uncertainty *qua* ambivalence is important in itself, I will explore it in detail in 4.2. When exploring the topic of diagnostic uncertainty later, I will say more about its contribution to misdiagnosis. For now, let us focus on the second potential source of misdiagnosis, which is the *lack of relevant information*.

---

haviour had no negative consequences for the patient does not seem make their behaviour less problematic and unprofessional considering what should be expected of a clinician. To make an intuitive comparison: whether a driver engaged in speeding should be determined not by the consequences of them speeding, like hitting someone or not (although this might be relevant in court), but by what constitutes speeding and whether the driver did what we consider to be speeding. If you disagree, this is no problem; nothing really depends on this preference of mine. If you do disagree, you could just add in the harmfulness condition on top, and the rest of my explanation in terms of the model-based account would not change.

Lack of relevant diagnostic information might come about in many ways. Patients might intentionally misinform or hold back information from the diagnostic expert, or they might misremember or have forgotten things when asked about them. They might have performed intentionally badly in cognitive tests, or just have been unmotivated to cooperate and therefore not performed well. Or they might simply misunderstand the instructions or questions but appear so confident and competent that the clinician had no reason to think that there was a problem.

Imagine a patient showing the objective complaint of reluctant speech behaviour. As discussed in the last chapter, such speech behaviour may point towards the psychiatric symptom of *alogia* and so is of interest to the psychiatrist. As we also discussed in the last chapter, besides being alogia, reluctant speech might occur as a medical symptom in the context of a traumatic brain injury, or the patient's speech behaviour might result from the patient's intention to be uncooperative. Let's say that the patient intended to be uncooperative – specifically, to make the psychiatrist think they had a traumatic brain injury. If the psychiatrist interviewed the patient to gather information in order to evaluate the models for the respective diagnostic options, the patient could simply pretend to be unable to give longer answers if required and could say that he has not always been like this, which would be supported by relatives and friends of the patient because he indeed is not normally like this. This would then exclude the model for the diagnosis of motivated monosyllabism. Also, he would easily be able to pass the cognitive tests evaluating the presence of alogia discussed in the last chapter. Finally, the patient might then claim to have stumbled over a chair today, hit his head, briefly lost consciousness, and has the feeling that he lost some time afterwards. He may claim that he felt disoriented for a minute after this and was feeling sick. Maybe this patient planning the fraud even hit himself with a stick, hard enough to have a bump on his head to support the illusion. Although a CT scan provided for the patient would not show any lesions, the rest of the story and the overall evidence would perfectly fit the case of a traumatic brain injury, and not every traumatic brain injury necessarily shows up as a lesion in a CT scan of the brain. In conclusion, the psychiatrist would likely and wrongly conclude that the complaint of the patient's reluctant speech results from a traumatic brain injury. This wrong conclusion, however, would be a misdiagnosis rather than malpractice, because at this point the psychiatrist invested reasonable effort and carried out the required diagnostic procedures to gather the diagnostically relevant information, but arrived at a wrong conclusion based on an informational bias. This bias did not result from the psychiatrist doing anything that would go against good diagnostic practice guidelines, and so we would usually not consider him to be at fault for having arrived at this wrong conclusion. So much for misdiagnosis for now; we will return to it in 4.3. Now let us turn to what would constitute a case of wrong diagnosis *qua* malpractice.

As in the case of misdiagnosis, let me point out what may go wrong in the diagnostic process as presented by the model-based account in the case of malpractice. While misdiagnosis occurs when all steps are carried out correctly but there is a residual uncertainty or misleading diagnostic information that leads to wrong diagnostic conclusions, malpractice occurs if the psychiatrist makes significant mistakes in the procedure of diagnostic reasoning. Again, this procedure consists in listening to the spontaneous complaints of the patient and systematically evaluating their psychopathological status; considering the various potential models of psychopathological, other medical, or non-medical conditions; testing and examining the patient for information relevant to evaluating these models against the patient's presentation; selecting the best-fitting models for the present complaints; providing a formulation based on the selected models; and providing one or more diagnosis based on the classification rules of the manual in use and the symptoms identified in the case formulation. In any of these steps, the psychiatrist could make mistakes leading to a wrong diagnosis, constituting a case of malpractice. Here are some examples. Psychiatrists might not spend enough time listening to their patients' complaints, or might incompletely assess their mental status, which then leads them to fail to consider all relevant models and therefore to end up not evaluating all relevant complaints. They might make mistakes in selecting a best-fitting model for patients' complaints, because they do not invest enough effort in thinking about which model is best supported by the information gathered about the patient. Or they might not pay close enough attention to the diagnostic criteria of disorder diagnosis and provide an unjustified diagnosis. In all these cases, the psychiatrist would be at fault for the wrong diagnosis and the harm that might take place in consequence of a wrong diagnosis produced by malpractice, because they did not fulfil their diagnostic responsibility at the level of the diagnostic procedure.

Taking this approach to misdiagnosis and malpractice, what does it do to help us identify and distinguish between them? Imagine an instance in which a patient has received a diagnosis that has later been judged to be wrong, and that this patient has received treatment based on this diagnosis that was harmful – for instance, because of side-effects of medication that she would not have been prescribed if her initial diagnosis had been correct. Now the patient is pressing malpractice charges against the practitioner. For someone to decide whether the wrong diagnosis of the patient resulted from malpractice, rendering the clinician at fault, or was a misdiagnosis that is not the fault of the clinician, someone investigating the case would have to answer a question deriving from the most general understanding of malpractice and misdiagnosis, as presented in the first paragraph of this section: did the wrong diagnosis result from the practitioner not carrying out the diagnostic procedure with thoroughness, or because the diagnosis was based on wrong or incomplete information, or on information that led to diagnostic ambivalence in which the wrong choice appeared plausible? The interpretation of the difference between malpractice and misdiagnosis in light of the model-based account to diagnostic reasoning provides an approach to answering this question in principle.

If there is sufficient information available about the diagnostic process that was carried out and the diagnostic considerations made by the diagnostic expert (e.g., in the form of documentation, notes, the case formulation, and (honest) reports), someone investigating the charge of malpractice may look at this information to evaluate whether it indicates that the clinician followed the model-based diagnostic reasoning step by step in the way outlined earlier and presented in detail in the preceding chapters. If not, this would suggest that the clinician engaged in malpractice. If no malpractice took place, the only other option is that the wrong diagnosis is classified as a misdiagnosis. If, however, the investigation comes to the conclusion that somewhere in the diagnostic process malpractice took place and led to the wrong diagnostic outcome, the clinician will be responsible for the wrong diagnosis and the consequences of actions that were taken or not taken based on it.[3] In this way, the model-based account helps us to differentiate and identify instances of misdiagnosis and diagnostic malpractice.

---

3 It could be the case that although some aspect of the diagnostic process qualifies as malpractice, correctly carrying out the diagnostic procedure would have made no difference. In other words, the same wrong conclusion would have been drawn even if no malpractice had taken place. This might happen, for example, because in another part of the diagnostic process an important piece of information was not accessible to the clinician even though everything was done right in this part of the diagnostic procedure, while the part of the diagnostic process that was carried out wrongly would not have provided information or conclusions that would have made a difference. For example, it might be that the clinician did not carry out a proper mental status examination but did not miss anything relevant to the wrongly made or potential correct diagnosis because of this. It was a patient's lie later in the interview that led to the wrong evaluation of a complaint as some particular symptom and in the end to a wrong overall diagnosis – as, for example, in the case of the patient faking the TBI. In this case, malpractice took place but this malpractice would not be the cause of the harm to the patient. This again may have different legal consequences and depending on our moral stance might also make moral differences. Malpractice took place nonetheless. And again, the model-based understanding provides the resources for deciding whether the malpractice is responsible for a potential harmful outcome. It can help us evaluate where in the process specific diagnostic decisions have been made in the context of the evaluation of diagnostic models against diagnostic information, and so can tell us which step in the process was relevant to which conclusion. If, given the analysis of the diagnostic process that took place, no lack of information, misused models, or inferential mistakes resulting from the malpractice in this case seems to be responsible for the wrong diagnostic choice, the wrong diagnosis would be a misdiagnosis even though there was also malpractice involved in the overall diagnostic procedure.

4.4 Diagnostic Uncertainty through Ambivalence

Another phenomenon well known in clinical contexts is diagnostic uncertainty and the attempts to overcome it. While it is sometimes easy to determine what the diagnosis of a patient should be, this is not always the case. There are occasions on which psychiatrists are uncertain about diagnostic decisions because what they have learned about the patient seems to allow for several potential diagnostic conclusions, so that additional effort is necessary to carve out which among the plausible diagnostic options might be the best. And even then, finding a certain answer might not always be possible. How uncertainties in diagnostics arise, and how they might successfully or unsuccessfully be resolved, will be the focus of discussion in this section. In addition, I will say a few words about how, despite great effort, a failure to resolve uncertainty might lead a psychiatrist to draw a wrong diagnostic conclusion, and why such cases are misdiagnosis rather than malpractice. This discussion will supplement the previous work in 4.3.

For psychiatric diagnostics we must consider two levels of uncertainty: the level of syndromal diagnosis and the level of symptoms. On the syndromal level, clinicians may be uncertain whether they should attribute a certain mental disorder diagnosis (X) to a patient or not, whether they should attribute one or another diagnosis (X or Y or ...) to a patient, or whether they should attribute more than one diagnosis (X and Y and ...) to a patient. Although this level of uncertainty often occurs, it is philosophically relatively uninteresting from the perspective of the model-based account, because how this decision must be made in accordance with best practice is solved by the major diagnostic manual in use, and if it were not solved by the manual, there would be no right or wrong way to do it.

In general, a diagnostic evaluation produces evidence of a sufficient standard to allow us to infer the presence of symptoms and so to provide a diagnosis whose list of diagnostic requirements most closely matches the patient's presentation, maximising the number of psychopathological relevant features addressed by one diagnosis. Whether a subset of the diagnostic features already employed to provide this diagnosis is allowed to be used again to justify another diagnosis is case-dependent. The DSM-5 (APA, 2013, pp. 155f.), for example, does not intend clinicians to reuse symptoms used to diagnose a major depression to additionally diagnose a patient with moderate and mild depression. However, it does allow clinicians to reuse them to additionally diagnose patients with dysthymia (ibid., p. 168), which would be what is usually called a double depression. The DSM-5 does support diagnosing agoraphobia (ibid., p. 218) on top of a panic disorder (ibid., pp. 208f), but not panic disorder if panic attacks occur in response to social situations (i.e., social anxiety) (ibid., p. 209). Manuals also offer many diagnostic options to account for leftover symptoms that are insufficient to support an independent diagnosis. The DSM-5 (ibid., pp. 160f.), for example, allows us to specify that a major depression diagnosis is accompanied by anxiety features that in themselves do not suffice for an anxiety disorder diagnosis, by adding the specifier "with anxious distress" to the diagnosis (ibid., p. 161). And finally, for certain disorders that are clearly approximated in terms of present symptoms but not fully met by the diagnostic findings, there are diagnostic categories that allow clinicians to classify these as well. For example, according to DSM-5, cases in which several depressive symptoms are present but no constellation is observed that would allow for any formal diagnosis of depression, the clinician is supposed to diagnose "other specified depressive disorder", which is a "presentation whose symptoms [are] characteristic of a depressive disorder that causeses clinical significant stress […] but do not meet the full criteria for any of the disorders" (ibid, p. 165). Whatever critique we might wish to make of the major diagnostic manuals DSM or ICD from the perspective of the model-based account – which, remember, is not an attempt to criticise diagnostic practices but rather an effort to make them intelligible – it does not seem that if well applied, these manuals leave the diagnostic expert who is aware of the symptoms of their patients in the dark about what diagnostic decisions they have to make. However, the "who is aware of the symptoms" qualifier brings us to the philosophically more interesting instances of diagnostic uncertainty from the model-based perspective: uncertainty regarding what symptom to attribute.

Diagnostic uncertainty regarding symptoms can occur in various patterns if it is not unequivocal which symptom value an initial complaint should be assigned after the patient has gone through the diagnostic process. The psychiatrist might be uncertain as to whether a complaint should be evaluated as one psychiatric symptom or another, or as a medical problem or a non-medically relevant issue instead. Such uncertainty often occurs in clinical contexts and may force the clinician to think harder or do additional diagnostic work to reach a solution, which sometimes but not always works. Uncertainty may persist as to whether a patient's complaint clearly qualifies as a psychiatric symptom or is a psychological complaint of non-clinical value. How exactly we can understand the occurrence of such uncertainty and the ways in which it may be resolved? Here is how the model-based approach can account for it.

If we consider the above-described diagnostic uncertainty regarding symptoms via the modelling account, it appears there are three possibilities for how it may arise:

i) None of the models set up for an initially recognised complaint matches the patient's well enough to be accepted. As a result, the psychiatrist has no unambiguous basis on which to make any judgement for or against evaluating the complaint to be a psychiatric symptom, a medical complaint, or a non-medical issue.

ii) More than one model for a complaint from amongst those set up based on knowledge from the domain of psychiatry (e.g., models that would render the complaint psychiatric symptom (a) or (b)) fits the patient's condition sufficiently well to be accepted. As a result, the psychiatrist has no unambiguous basis on which to make a diagnostic judgement regarding the initial complaint.
iii) More than one model for a complaint from amongst those set up based on knowledge from a range of domains (i.e., psychiatry versus other medical or non-medical fields) fits the patient condition sufficiently well to be accepted. As a result, the psychiatrist has no unambiguous basis on which to make a judgement for or against evaluating the complaint to be a psychiatric symptom.[4]

In all these cases, the decisions regarding the psychiatric symptom value of a complaint cannot simply be looked up. If we have only the complaint as the prior, there is no straightforward formal way to derive the correct evaluation in the way we can do it if we are on the level of disorder diagnostics, already equipped with a set of symptoms that we can take as priors to decide which disorder(s) to diagnose. How, then, do we overcome such a situation? The psychiatrist has several options. Some of these options are attempts to deal with the uncertainty by forms of further theorising and evaluation, while others present pragmatic solutions. I will discuss in turn the three instances of uncertainty and how they can be addressed by such means.

The first type of uncertainty, resulting from no diagnostic model suiting the patient's presentation sufficiently well according to the fidelity criteria assumed for the tested models, is the most severe case of diagnostic uncertainty. Think of an example of a patient reporting anxiety. On close evaluation, it turns out that this patient does not show any signs of the typical cognitive style and somatic reactions of anxiety that would allow the psychiatrist to identify their anxiety as a psychiatric problem. The patient has also had no recent experiences that would render his currently high anxiety level understandable. He has taken no medication and has no physical condition that might induce such reactions. The severity of such cases lies in the problem that there are no theoretical resources that seem to provide a theoretically justified diagnosis, because the complaint matches no diagnostic models whose application would justify the inference to any diagnostic conclusion regarding a complaint. The psychiatrist just has no way to say what is going on here, and ideally this would also become clear in the psychiatric case formulation.

---

4 What about the option of multiple medical but non-psychiatric models, or wholly non-medical models, fitting equally well? While this option exists, I will not discuss it here, as in these cases it is to be assumed that the complaint is not a psychiatric symptom and further diagnostic efforts would either be a matter for another medical profession (where multiple non-psychiatric medical options fit) or be of no medical interest at all (where multiple non-medical models fit).

Pragmatically speaking, a psychiatrist may nonetheless support the evaluation of a complaint as a symptom or a medical problem initially suggested by the complaint. In such a case, the clinician would end up making what has been called a *suspicion diagnosis*. A suspicion diagnosis may be understood as the diagnostic proposal that is the most plausible option given all diagnostic evidence but that is still not sufficiently certain to fully endorse it. It is supported by pragmatic considerations regarding the cost/benefit calculus of treating a patient according to this diagnosis versus another diagnosis versus refraining from providing any diagnosis and not treating the patient at all.

To give an example, it might be the case that a patient meets all but one criterion sufficient for a major depressive disorder (MDD) and displays a complaint that, if it were a symptom. would allow for this diagnosis. However, no model evaluated suggests that the complaint be considered a symptom. Further, imagine that there is a certain intervention that, based on treatment guidelines, is intended to be provided only to MDD patients, but there is a good chance that this intervention might help the considered-close-to-MDD patient, because there is some evidence that it may help reduce symptoms in other depressed but not MDD patients. In such situations, psychiatrists take the path of what has been discussed in the literature as "workarounds" (Whooley, 2010): they diagnose as if the complaint were a symptom. While everyone working in clinical practice will be familiar with such patterns of practical reasoning, the question of course arises as to whether these patterns of reasoning are rational and ethically permissible considering the overall practical purpose of psychiatry to help patients, or whether other considerations (e.g., the risk of biasing epidemiological studies based on clinical data, not meeting general standards of evidence-based practice) speaks against such practice. I will remain agnostic regarding this normative question.[5] To come back to our anxiety example, the psychiatrist may for pragmatic reasons decide to consider the initial complaint of anxiety as a psychiatric symptom for the practical purpose that this might allow for a diagnosis that could be used to justify therapeutic or pharmacological treatment, so that there is at least a chance of improving the patient's condition.

The second and third type of uncertainty occur if there are several models of a psychiatric complaint that match the patient's presentation sufficiently well, while at least one of these model, if chosen, would render the complaint a psychiatric

5 The pragmatic reasoning process feeding into suspicion-diagnostic conclusions is a kind of clinical reasoning rather than diagnostic reasoning. The interaction of this clinical reasoning with theoretical diagnostic reasoning evaluating the initial plausibility of diagnostic conclusions purely on the basis of diagnostics is an interesting and clinically relevant topic. However, delving into the logic of pragmatic reasoning in clinical diagnostics would require a new line of investigation and is thus beyond the scope of my project, which focuses on epistemic (i.e., diagnostic not clinical) reasoning. I will therefore not discuss the topic of how exactly suspicion diagnosis is provided and justified, but only outline its structure and purpose.

symptom. Going back to the anxiety example, a patient reporting the complaint of anxiety might present in the in-depth evaluation such that a model evaluating the anxiety as a psychiatric symptom – by assuming a model of anxiety's typical cognitive style (including attentional bias, memory bias, and interpretation bias) – applies sufficiently well. At the same time, a model that assumes the anxiety to be a normal psychological reaction in light of a model assuming a combination of environmental factors to increase stress in the patient, making their anxiety response normal, also fits the patient sufficiently well. That is, it appears justified to assume the patient's complaint to be a psychiatric symptom as well as a normal psychological reaction.

To resolve uncertainty in this instance, two approaches seem to be available. For a theoretical solution, the matching models may be compared in terms of how good their match is with the targeted complaint of the patient. If it turns out that one model matches the patient's presentation better than the other model, even though both models seem to be in principle applicable, it appears rational to choose the best-fitting model to make a diagnostic decision as to how to classify the patient's complaint. If, for example, two propositional models target the same complaint and from each model enough central propositions apply to the patient's presentation that in principle both models seem to match the patient's presentations, the diagnostic expert will go for the model that contains more diagnostic propositions that match with the patient's presentation – that is, the model that is a better fit. Of course, the judgement of "better fit" again has its complexities. Typical goodness-of-fit models that can be used in mathematical modelling to quantify how well a model matches with observations of the modelled system, producing a numerical value that allows for a decision between models, do not seem straightforwardly applicable given that we are dealing with qualitative models. Rather, it appears useful to ask what fraction of the total number of the propositions that the models consist in, beyond those sufficient to make a well-fitting candidate, are met.[6]

If this procedure does not lead to a conclusion favouring one model over another, because again both models seem to apply equally well, uncertainty is residual. Then the clinician must either refrain from drawing a diagnostic conclusion regarding the

6 Here, another weak point of psychiatric modelling (beyond its potential vagueness due to its qualitative format) surfaces. Since the models used to identify psychiatric symptoms are constitutive models, they do not necessarily entail any claims about specific causal relationships or aetiologies of the phenomena they attempt to model. They only identify constituents that must be present to attribute a symptom. The problem with this account is that if the constituents of more than one type of model apply equally well (or at least indistinguishably similarly well), to decide between them becomes impossible. What could solve this problem would be evaluating which potentially constitutive features are also causally responsible for the patient's presentation. An option that is not at its disposal of psychiatric modeling as it stands. Coming up with reliable causal models that would allow us to evaluate psychiatric symptoms would be beneficial in this regard.

complaint or opt for the pragmatic solution strategy, assuming an evaluation without fully endorsing it in order to support a suspicion diagnosis as described above. However, it seems that in this context a suspicion diagnosis, although still not unequivocally supported by evidence, would be epistemically stronger, because there is at least some evidence base that in principle would be sufficient to support the diagnosis, rather than no evidence speaking for it. The pragmatic decision could therefore be made with a higher base level of confidence and perhaps with fewer alternatives that are equably plausible compared with cases where no model seems to match the complaint, and where all models are similarly (un)likely. As a result, however, the diagnosis of the symptom may be wrong, and its suspicion-diagnostic support may allow for a syndromal diagnosis that is wrong. Yet after all the diagnostic steps have been carried out correctly, arriving at such a diagnosis for pragmatic reasons, such as allowing for a most plausible and least harmful treatment that might potentially improve the patient's condition, is in line with the pragmatic aims of psychiatry to cure and care for patients. And if the conclusion turned out to be wrong, this would make it a misdiagnosis rather than a case of malpractice. In this way, I have also outlined the missing way to arrive at misdiagnosis, as promised in the previous section. Next, let us turn to the topic of instinctual diagnosis.

4.5 "Instinctual" Diagnosis

If one works in clinical context, say a psychiatric hospital, a story like the following will perhaps be familiar. A senior physician is coming to see a new patient who just got admitted to the psychiatric unit. She enters the room and exchanges only a few words with the patient. She then leaves the room and says to her colleagues something like "I suspect the patient has an XYZ diagnosis". And it turns out after more detailed diagnostic procedures that the senior physician was right. It seems that she has a special diagnostic "instinct". How can we explain how such often reliable instinct works, what its epistemic benefits and downsides are, and why we apparently want the actual diagnosis to be made according to formal standards even if we have a clinician with great intuition around? The model-based approach provides us with a story that allows for a plausible approach to all these questions.

Let us go back to the situation of the short encounter between a clinician and a client from which such an instinctual diagnosis might result. What is going on here? Plausibly, in a short encounter with a patient, the psychiatrist will at best be able to become aware from observation or incomplete evaluation of a limited number of complaints of the patient. Although no full picture of the patient's complaints can be claimed, since no complete screening has been conducted, the physician will at least have gathered some information about the most salient complaints of the patient, though not the necessary information to evaluate them properly for their symptom value. In other words, the clinician has conducted an incomplete first step in the proper diagnostic process. What is she doing with the information to arrive at a diagnostic conclusion? The spotted complaints are treated *as if* they would have turned out to be psychiatric symptoms. The psychiatrist has a list of potentially present list of symptoms to hand and can think through the limited number of disorders that would match with this pattern, proposing that the patient will perhaps suffer from the disorder(s) matching the assumed symptoms that are most likely present, possibly for a subset of the clinical population that the patient falls into on first glance (e.g., as regards sex or age).

Although such quick likelihood assessment may generate a first hypothesis as to what might be the patient's disorder that may turn out to be correct, this approach to diagnosis often has the problem that it is not comprehensive or supported by evidence. In diagnosing a patient, we expect diagnosis to be supported by the best available evidence that can be collected with reasonable effort to determine what the patient's problem may be, so that they can be offered the most beneficial treatment for their condition and we can avoid harming them by offering wrong treatment or withholding better treatment options from them. In this case, there is a fair chance that we will do exactly this, since we cannot know whether any of the complaints would indeed be evaluated as psychiatric symptoms if properly assessed. A complaint may not be the symptom of relevance and may therefore mislead the diagnostic guess. There is also a risk that it is such a symptom but that this symptom is not part of the most likely psychiatric syndrome, or that the pronounced symptom is present but not enough other diagnostic criteria are met in addition to it to diagnose the suspected condition. Also, complaints that were not picked up on by a short encounter will not be considered in the diagnostic guess, and these might have pointed towards highly relevant symptoms that would have led to a different diagnostic conclusion. Hence, basing one's diagnosis on a short encounter and a diagnostic guess seems to harbour a significant epistemic risk of being wrong. As being wrong in this case would mean being wrong because of a lack of proper diagnostic procedures, taking this risk and ending up with a wrong conclusion would indeed mean having engaged in malpractice, which is why usually "instinctual diagnosis", although it provides some guidance for a clinician to think about what might be wrong with their client, is not accepted as a proper approach to diagnosing patients.

In the above case, we assumed that the diagnostic guess was the most rational possible based on the best knowledge of the likelihood of symptoms and disorders in certain reference populations of patients, under the assumption that every spotted complaint would be a psychiatric symptom. Another problem arises if we bear in mind that humans, especially when they think rapidly, are anything but perfect rational machines. In rapid diagnostic decisions, humans tend to unintentionally apply heuristics that bias their decisions (Tversky and Kahneman, 1974). Heuristics that are important in diagnostic contexts appear to be, for example, the availability heuristic (which leads us to judge how frequent or probable something is based on how easily we can bring to mind an example of a state of affairs, leading us to mistake actual availability for actual frequency) and the representativeness heuristic (in which we assume that someone belongs to a category because they seem to match the stereotype of this category) (Tversky and Kahneman, 1981). Both are found to be widely present in expert judgements, including in the diagnostic judgements of medical and psychiatric experts (e.g. Elstein, 1999; Garb, 1996; Koehler, Brenner, and Griffin, 2002; Ægisdóttir et al., 2006). Therefore, on top of the likelihood of being wrong in an "instinctual diagnosis" even if we were perfectly rational and well informed, our own human psychology is an additional problem. Our psychology might bias us to judge patients as falling into one or another diagnostic category just because we as clinicians happened to see patients showing a certain complaint as matching a stereotype of someone having a certain disorder, or because in the limited sample size of patients we have seen, patients with a certain complaint mostly turned out to have this disorder.

As an example, think of a patient who is harming himself without the intent to kill himself. Such behaviour may indicate the psychiatric symptom of nonsuicidal self-injury (NSSI) (Klonsky, Victor, and Saffer, 2014). NSSI is present, for example, in autism spectrum disorder (Johnson and Meyers, 2007), borderline personality disorder (Oumaya et al., 2008), bipolar disorder and dissociative disorders (Joyce et al., 2010), eating disorders (Rodríguez-López et al., 2021), depression, phobias, and schizophrenia (Singhal et al., 2014), non-suicidal self-injury disorder (Zetterqvist, 2015), and Munchhausen syndrome (Humphries, 1988). Looking at the available data, we learn that patients admitted to psychiatric hospitals with self-harm seem to suffer most frequently from depression or anxiety or and alcohol misuse, as well as attention deficit hyperactivity disorder (ADHD) and conduct disorder in younger individuals (Hawton et al., 2013). As pointed out by Hawton et al. "[t]hese findings are clearly at odds with the commonly held but misinformed view that the majority of self-harm patients do not have psychiatric disorders, or if they do then this is most likely to be a personality disorder." (Hawton et. al. 2013, p. 828).

However, there are also reasons for self-harming reported in the literature that do not seem to point towards psychopathology, such as religious reasons or the requirement to do so to be part of a certain subculture (Edmondson, Brennan and House, 2016). If a psychiatrist, knowing all this, briefly encounters a patient showing signs of self-harm or reporting having harmed himself, the first idea that springs to mind might be that this patient suffers from those disorders most frequently associated with this behaviour if it is a psychiatric symptom, and often enough the psychiatrist will be correct in their guess. However, in many cases this guess might also go wrong. Considering the example of self-harm, the patient may suffer from a different mental disorder associated with the suspected symptom(s) assumed based on the complaints (e.g., non-suicidal self-injury disorder rather than borderline personality disorder). Or the behaviour may not be a symptom of a mental disorder but rather a religious practice. Moreover, if we do not assume a perfectly informed and rational clinician but one whose decisions are potentially biased by availability and representativeness heuristics, the clinician might, after assuming the patient's behaviour to be a self-harm symptom, even more rapidly come to the conclusion that the patient suffers from borderline personality disorder. This might happen if the clinician worked for years in a hospital unit specialised in treating borderline personality patients who often showed this behaviour, so that there is now a tendency to equate self-harm as a symptom with the presence of a borderline personality disorder.

Looking at this example, it becomes clear why no responsible trained clinician should base their final diagnostic conclusions on their instinctual or educated diagnostic guesses. Thorough evaluation of diagnostic models against patients' presentations based on proper diagnostic information provides a better justification base for diagnostic conclusions than the above-described likelihood judgements. It does so because evaluating what indeed is the situation with a patient and matching this with our best psychopathological understanding of what is constitutive for a present psychopathological symptom tells us what is the case with the patient, rather than only telling us what the case with the patient might potentially be with a certain probability if a certain model fitted the patient. By following the proper process, the diagnosis also achieves diagnostic superiority, because if it is based on the process of model evaluation, it is supported by evidence that allows the inference of the presence of a certain symptom to be an inference to the best explanation. This inference occurs via the acceptance of a constitutive model that provides a constitutive explanation of how to understand the patient's complaint.

To avoid obviously problematic approaches by which diagnostic conclusions like the one discussed in this section may be reached, and also to make sure that there are no smaller mistakes in the process of diagnostics, there is an important tool at our disposal: critical diagnostic reasoning – that is, the critical diagnostic examination of one's own and others diagnostic work. This form of critical engagement with diagnostics is the topic of the next section.

4.6 Diagnostic Disagreement

Clinicians can be wrong about their diagnostic proposals for various reasons, some of which we explored above when we talked about misdiagnosis, malpractice, and diagnostic instinct. Knowing all too well that diagnostics is fallible, it is generally considered important to ensure that as many mistakes as possible are prevented or at least corrected.

Good clinicians try to do this with their own diagnostic conclusions once they have arrived at them by putting their own proposal and the way they arrived at it to the test again. If, after their self-assesment process, they still support their diagnosis, they will also evaluate it again later if interventions lead to changes that may require a diagnostic re-evaluation, or if any additional diagnostically relevant information is obtained that might require correcting their initial diagnostic judgements. But self-monitoring is not the only thing that happens. Besides monitoring their own work, clinicians also monitor each other if they disagree with a diagnostic conclusion and discuss this disagreement with each other, or at least they may ask colleagues to explain the reasoning behind a certain diagnostic conclusion – something that takes place in particular between new clinicians and their supervisors, to assess and train their diagnostic reasoning. Engaging in this kind of self-criticism and intrapersonal criticism of diagnostic decisions and resolving differences between two mutually exclusive evaluations is called critical diagnostic reasoning. This is thought to be an important feature of diagnostic reasoning as practised by clinicians, no matter their specialisation (Harjai and Tiwari, 2009; Mamede, Schmidt, and Rikers, 2007).

To engage in critical diagnostic reasoning, clinicians ask themselves or others questions that make them check their diagnostic decisions. For example, "Why exactly did I/you draw this diagnostic conclusion?", "What could be an alternative explanation?", "Did I/you consider all available and potentially relevant information?". Answering these questions by presenting a valid inferential path leading to the diagnosis, in support of which relevant information was gathered and adequately considered, can support one's confidence in one's diagnostic judgement, or, if the answers hint at flaws, undermine it. Alternatively, if there is a disagreement between clinicians, answering this question on both sides of the conflict and demonstrating how the diagnostic reasoning process on each side meets or fails to respond to these questions may lead to a rational agreement as to whether one or the other or maybe neither option seems to be right, or whether there is a residual uncertainty about whose the right diagnosis is. Now, how does the model-based account make sense of these intra- and interpersonal procedures?

Intra or interpersonal critical diagnostic reasoning is structurally equivalent to the procedures that can be employed in the case of diagnostic uncertainty discussed earlier. Therefore, the relevant points are quickly made. At the top level of syndromal diagnostics, the model-based account has nothing particularly interesting to say beyond what is to be found in the diagnostic manuals considering disorders to be sets of symptoms and using additional criteria to tell us straightforwardly whether a diagnosis is correct or not. Critical reasoning on this level simply requires double-checking whether all diagnostic criteria have indeed been met. This may be done for oneself (intrapersonal) or between clinicians (interpersonal). And again, it is the symptomatic level that seems to be more interesting. In other words, while there is little to no room for disagreement about what must be present for a major depression, because we can look it up in the manual we are using, whether the required symptoms are present (i.e., whether a patient's report that he no longer has fun when pursuing his hobbies is indeed a case of anhedonia) offers a livelier ground for diagnostic disagreement.

Critically evaluating whether attributing or not attributing any specific psychiatric symptom is adequate provides more room for the application of the model-based diagnostic reasoning framework. Its application is in principle like the method discussed earlier in cases of diagnostic uncertainty, since doubting the adequacy of one's diagnostic decision basically amounts to intentionally introducing artificial uncertainty. If a clinician is coming back to a diagnostic evaluation of a complaint, they may ask themselves whether they did carry out the initial evaluation (screening) of the patient in a way that covered all relevant areas, whether they considered the models for all encountered complaints, whether they considered all models relevant to the encountered complaints, whether they did what was required to generate data that allowed for the evaluation of the relevant models in the in-depth evaluation, and whether as a result of the comparison they chose the right model to apply.

The same may take place on an interpersonal level. Here, the debate between clinicians may start from various points. A supervisor or chief may want to discuss a diagnostic conclusion of a trainee to test and exercise their diagnostic reasoning skills based on a patient case that the supervisor themselves has never seen. Or a debate might result from a chief physician reading the case formulation supporting the chosen syndromal diagnosis of a patient but being unsatisfied with the justification provided by it. Or maybe colleagues in a team end up disagreeing about a diagnosis of a patient they are treating together and have to sort out this disagreement. In any of these cases, the clinician whose diagnostic conclusion on the level of symptom attribution is in question will have to make transparent the actions undertaken to gather initial and additional information about the patient, the models considered to apply, and why each model based on detailed diagnostic information was accepted or rejected. Making transparent this process then opens the field for interpersonal criticism. The colleagues or supervisors may point out that some models were not considered or sufficiently evaluated, suggest that the diagnostic data were insufficient to assume that one of the tested propositional models indeed applies to the patient, or raise many other points regarding any stage of the diagnostic process. If the interpersonal disagreement comes to a point where both debaters agree that each other's diagnostic evaluation is in principle valid, they might nonetheless think that their diagnostic choice is to be preferred because the model they picked better suits the patient's case. This situation may then be debated further, considering the theoretical solution strategy for diagnostic ambivalence earlier in this chapter, with the same potential outcomes: a solution in favour of one diagnostic conclusion or a residual uncertainty. To sum up: the way in which we can understand the occurrence of diagnostic disagreement and critical diagnostic reasoning in the context of psychiatric diagnostic reasoning is well covered with the resources of the model-based account.[7]

4.7 Change in Diagnostics

The final topic I wish to cover in this chapter concerns how an answer to the Methodological Question is capable of making sense of the possibility and limits of integrating changes into our understanding of psychopathology and the means we use to assess it. That an answer to the Methodological Question should have something to say about this is desirable for at least two reasons. First, because a good answer should be able to show that it will be able to assimilate modest changes in our understanding of psychopathology and methods of assessment. Small to modest changes occur all the time, and for an answer to provide a somewhat stable proposal that applies to psychiatric diagnostic reasoning at least in the recent past and will probably apply in the near future, it should be flexible enough to incorporate such changes. Second, it is important because only if the proposal can display its limits on implementing changes will it appear to be usefully precise. If significant changes that we could imagine taking place in a potential or fictional future of psychiatric research could be accommodated by the proposal without problem, it would seem too arbitrary to be considered a specific understanding of the diagnostic practices at hand.

In the following, I want to show how the model-based approach holds up to both requirements. To show the robustness of my account against small to modest changes but its sensitivity to relevant changes in psychopathology, I will discuss aspects of the two levels of diagnostics. The higher level of diagnostic decision-making will be discussed in terms of providing a symptom-based syndromal diagnosis,

7 What has been discussed in the previous sections on instinctual diagnosis and diagnostic disagreement, especially intrapersonal diagnostic disagreement, can also be found under discussion – sometimes in normative terms, sometimes in descriptive terms – in the medical education science literature on diagnostic reasoning. The error-proneness of quick and intuitive judgements and the relevance of analytic reasoning as their corrective have been discussed in the context of dual-process theories. These theories consider human cognition to consist of two interrelated systems, one of them intuitive, the other one analytic, with the intuitive being more prone to several kinds of bias (Monteiro and Norman, 2013). Applications of this idea in medical education assume that the same is true for diagnostic reasoning: quick intuitive judgements pay the price of being open to all sorts of biases, such that any judgement made in this way (if one is using this approach at all) requires the monitoring influence of analytic reasoning (Croskerry, 2009; Elstein, 2009; Marcum, 2012).

and the lower level of psychiatric diagnostics will be dealt with in relation to psychiatrists' evaluations of the presence of symptoms. Considering the case where the lower level remains the same and only the top level is changed, I will discuss what the changes may look like such that the model-based account may still be useful to understanding psychiatric diagnostic reasoning, and also under which circumstances it may no longer be useful. Then, for the lower level of symptoms, I will look at potential changes in the understanding of symptoms by homing in on the symptom of anhedonia. I will first discuss varying historical understandings and the current understanding of this symptom. I will argue that the variations in these understandings, though real, is small enough in its relevance to how the symptom would be evaluated that adopting each version of it would square with the model-based account's understanding of symptom evaluation. This argument will demonstrate the flexibility of this level of the model-based proposal for clinical diagnostics.

Next, I will discuss the current science of anhedonia falling within the field of computational psychiatry and how it is changing our understanding of anhedonia. Although the changes in our understanding of mental symptoms like anhedonia that computational psychiatry is currently encouraging have not yet led to widely adopted change in the clinical evaluation, this may happen in the future. I will therefore discuss, mainly using the example of anhedonia, some of the options for how computational psychiatry may soon change diagnostic evaluations and point out which changes would not, but also those that would, undermine the model-based approach. This will demonstrate the fallibility of my approach in light of more significant changes in diagnostics on this level. Finally, I will provide a brief discussion of some possible though perhaps unlikely changes to psychiatric diagnostics that would significantly transform our understanding of both levels of diagnostics. I will argue that these significant changes would render the model-based account a chapter in the history of psychiatric diagnostic reasoning rather than part of its present. I will conclude that the model-based account is flexible and thus robust enough, but at the same time sensitive and thus fallible enough, to fulfil the desideratum in question. Let me begin by discussing the current format of syndromal diagnostics and how its changes might or might not affect the plausibility of my answer to the Methodological Question.

If we look at the contemporary format of psychiatric diagnostics, which is based on syndromal diagnosis consisting of clusters of symptoms and signs, changes may appear on two levels: either on the higher syndromal level or on the lower symptom level. On both levels there may be changes. Let us talk about the higher level first. Changes on this level may entail, and have entailed, new diagnostic categories such as the gaming disorder introduced in ICD-11 (Aarseth et al., 2017). The criteria for existing diagnoses may be changed, as occurred with the criteria for PTSD from ICD-10 to ICD-11 (Barbano et al., 2019). Or diagnostic categories might be abandoned, like the subtypes of schizophrenia in DSM-5 (Tandon et al., 2013), or introduced, like the subtypes of neurocognitive disorders in DSM-5 (Regier, Kuhl, and Kupfer, 2013).[8]

Although the central diagnostic manuals DSM and ICD may change in this manner, these and future changes of diagnostic taxonomy will not impact the ways in which these manuals are used as long as they keep operating in this framework of symptom-based syndrome diagnostics – that is, using identified symptoms and signs plus the additional diagnostic criteria to diagnose disorders. Accordingly, the symptom-based pattern recognition approach would perhaps not be influenced by these changes if the straightforward formal process of inferring syndromal diagnosis from patterns of symptoms remained the same. However, if the way in which we diagnose psychiatric disorders on the top level changed (i.e., if we still identified symptoms and signs but used them differently in a second step to make a higher-order diagnostic judgement), symptom-based pattern recognition approach might of course change too. To look at just one scenario that somewhat realistically might take place (or at least one that is argued for in the literature), namely that inferring disorder diagnoses as syndromes from specified clusters of necessary and sufficient sets of symptoms is no longer used, imagine that instead we only diagnose present symptoms. The rationale behind this could be, for example, that we can better target specific symptoms with specific interventions than syndromes that allow for very heterogenous clinical presentations under one label (Park et al., 2017). In this case, the overall model-based proposal would be no longer be correct but would contain superfluous components. Of course, superfluous components (i.e., everything that goes beyond symptom diagnostics) could be cut out to make the proposal adequate again, but for the time being it would be inadequate. This shows that my model-based proposal is in principle robust to some changes on the higher level of diagnostics (disorder diagnostics) but would also be open to falsification if deeper changes were to take place. Now we can move on to consideration how the model-based theory of diagnostic reasoning can handle changes in the context of the evaluation of symptoms and signs.

Whatever changes take place on a level of diagnostics higher than the level of symptoms – whether changes in the taxonomy of syndromes or a whole new way of making of attributed symptoms – they do not affect the way in which symptoms themselves are evaluated. However, there might also be changes in diagnostics that

8 Such past decisions regarding single changes in the diagnostic taxonomy, as well as the whole diagnostic approach of syndromal diagnosis based on symptom clusters (now supplemented with dimensional diagnostics of certain symptomatic features), have been heavily criticised by researchers, clinicians, and philosophers (e.g., Kendler and Parnas, 2012; Casey and Kelly, 2013; Demazeux & Singy, 2015; Hengartner and Lehmann, 2017; Ghaemi, 2018). But regardless of the validity of concrete categorisations of disorder entities, the delineations between them, or even the whole approach of syndromal diagnostics, diagnostic practice must apply it.

would influence the way we would identify symptoms. The way this may occur is through changes in how we understand these symptoms. Such changes in understanding may, on the one hand, lead to change regarding what we look for to evaluate the presence of a symptom by our usual means of diagnostic information-gathering and use, or it may be that our changed psychopathological understanding is accompanied by new means of evaluating the presence of a symptom. I will discuss both cases considering the model-based approach I have proposed, beginning by showing how the model-based account would accommodate for the first case: changing understanding with no general change of diagnostic approaches.

The idea in this case would be that our ways of grasping psychiatric symptoms *via* propositional models used to evaluate the presence of such symptoms, would change in so far as those propositions in the model change. However, despite modifying the model structure that we then use we would still follow similar process of screening, in-depth diagnostic information-gathering, and conclusion-drawing. To make this possibility more vivid, let us consider a historical example and ask how these different understandings would have been used in the context of temporary diagnostic reasoning as explained by the model-based account. Let us look at anhedonia.

As Berrios and Olivares (1995) point out in their historical investigation of anhedonia, we have seen many understandings of this symptom in the past hundred years or so. Although the phenomenon itself was described and discussed earlier, it was Ribot (1897, p. 53) who coined the term anhedonia and characterised it as a general inability to experience pleasure, found in individuals suffering from melancholia. Since then, anhedonia has been described clinically as present in patients suffering from depressive disorders as well as psychosis (especially schizophrenia) (Pelizza and Ferrari, 2009; Lambert et al., 2018).

Earlier discussion of ostensibly the same clinical phenomenon can be found in Griesinger (1861), calling it "mental anaesthesia": a state in which "the patient can no longer rejoice in anything, not even the most pleasing" (ibid., p. 223). Going into more detail, he described this phenomenon as a "continual dissatisfaction with the external world" and as involving "abnormal states of emotional dullness [Gemüthsstumpfheit], and even of total loss of emotions [volligen Gefühllosigkeit]" (ibid., pp. 66–67).

Later authors, not picking up on the term anhedonia, described the same phenomenon differently again. Kraepelin (1919, p. 33) wrote:

> The singular indifference of the patients towards their former emotional relations, the extinction of affection for relatives and friends, of satisfaction in their work and vocation, in recreation and pleasures, is not seldom the first and most striking symptom of the onset of disease (dementia praecox). The patients have no real joy

> in life, "no human feelings"; to them "nothing matters, everything is the same"; they feel "no grief and no joy", "their heart is not in what they say".

Jasper (1963, p. 93) talked about a clinically relevant "feeling of having lost feeling" (das Gefühl der Gefühllosigkeit) in which "patients complain that they no longer love their relatives, they feel indifferent to everything. Food does not gratify. [...] All sense of happiness has left them. They complain they cannot participate in things, they have no interest".

Myerson (1920) and others picked up on the term *anhedonia*. Myerson proposed an understanding of the phenomenon in light of a developmental model, summed up by Berrios and Olivares (1995, p. 463):

> [F]irst, by the disappearance or the impairment of the appetite for food and drink and failure in the corresponding satisfactions [...] Second, there is a failure in the drive or desire for activity and the corresponding satisfaction.... Third, the appetite or desire for rest and the satisfaction of recuperation are also involved in the anhedonic syndrome. The tired feeling [...] may be supplanted by a final absence of the feeling of fatigue.... Fourth, the sexual drives and satisfactions are conspicuously altered in the acquired anhedonic states. [...] Finally, the social desires and satisfactions, which belong indissolubly to the nature of the herd animal known as man, become disorganised, deficient and even destroyed.

Klein's (1974) understanding arguably went on to have the largest impact on the understanding of anhedonia that made its way into the DSM-III and later editions (De Fruyt, Sabbe, and Demyttenaere, 2020). He described anhedonia as "a sharp, unreactive, pervasive impairment of the capacity to experience pleasure or to respond effectively to the anticipation of pleasure" and as "a phasic, temporary, severe lack of present or anticipated satisfaction associated with the conviction that one cannot perform adequately" (Klein, 1974, p. 175). Later, Klein (1987) also added two dimensions to pleasure and its loss, distinguishing between consummatory pleasure, which is the pleasure of consuming or doing something that should be expected to bring pleasure, and appetitive pleasure, which is the pleasure gained from the expectation of a future usually pleasurable stimulus.

Considering this sample of historical views on what constitutes anhedonia as a symptom of mental disorder, linking those making similar proposals, and translating them into a propositional model would result in five different model: the Ribot model, the Griesinger model, the Kraepelin-Jasper model, the Meyerson model, and the Klein model. According to the Ribot model, the only proposition that would have to be shown to apply to a patient to justify the attribution of anhedonia is that the proposition "fully lacks the capacity for consummatory pleasure" applies to an individual. According to the Griesinger model, the propositions to apply to a patient would be that the patient has "dullness or loss of emotional reactions" and a "permanent state of dissatisfaction". The Kraepelin-Jaspers model would require the proposition "no expression or repot of emotional experience", "general indifference to occurrences in the surrounding world". The Meyerson model would require that the content of the following propositions apply to the patient and have arisen in the stated order: "loss of appetite and pleasure in food", "loss of drive for activity and the corresponding satisfaction", "loss of desire for and enjoyment of relaxation", "loss of sexual drive and satisfaction from sex", "loss of interest in and satisfaction from social interactions". And finally, the Klein model requires three propositions to apply, namely "loss of consummatory pleasure", "loss of anticipatory pleasure", and "believing that one would perform poorly in usually pleasant activities".[9] In contrast with these historically informed models we may also consider the diagnostic features of anhedonia in the DSM-5 text revision. Here, we have a list of features, where each of the features, separated by a comma, would make one proposition of the model:

> Feeling less interested in hobbies, not caring anymore, not feeling any enjoyment in activities that were previously considered pleasurable, reduction from previous levels of sexual interest of desire. Family members may notice social withdrawal or neglect of pleasurable avocations. (APA, 2022, p. 187)

Considering all these propositional models, including the current DSM-5 presentation, we can imagine how information sufficient to plausibly accept or reject the relevant propositions can be gathered by means of behavioural observation and interviewing of patients and conversations with relatives and friends (i.e., the typical current means of information-gathering), and therefore that while each of the models could in principle be adopted to determine the presence of anhedonia, all that would have to change for this would be the propositions to be evaluated in the otherwise similar diagnostic process. We would still use the same type of model and the same means of evaluation. This little look into the history of psychiatry therefore seems

9 Note that while all these models address anhedonia, they do not consider its occurrence in the context of the same disorder. Kraepelin's comments consider the occurence of anhedonia in dementia praecox (schizophrenia) while Klein describes anhedonia in the context of depression. Whether the psychiatric symptom of anhedonia in both patients is indeed the same across contexts which is usually assumed in the literature (e.g., Harvey et al., 2007; Pelizza and Ferrari, 2009) and also in the DSM-III, is challenged by more recent neuroscientific research. A better understanding of the neurobiology of anhedonia (Kuhlmann, Walter, and Schläpfer, 2013; De Fruyt, Sabbe, and Demyttenaere, 2020) begins to suggest that the cross-diagnostic symptom anhedonia may indeed represent two different conditions in the contexts of schizophrenia and depression. In depression, anhedonia may be characterised by impairments in anticipatory pleasure and integration of reward-related information, while anhedonia in schizophrenia is associated with neurocognitive deficits in representing the value of rewards (Lambert et al., 2018; Liang et al., 2022).

to support the idea that the model-based account shows some robustness, allowing us to integrate some changes on the level of symptom diagnostics and helping us to understand how they are integrated.

Instead of going down this very speculative path, I would like to bring up an example that seems more likely to be relevant to psychiatric diagnostics in the near future and see whether the methods accompanying it would necessarily or likely make the framework of model-based diagnostic reasoning obsolete. For this I will look at computational psychiatry.

Computational psychiatry as a field of research "consists of applying computational modelling and theoretical approaches to psychiatric questions" (Seriès, 2020, p. 12).[10] In this way, "Computational Psychiatry seeks to understand how and why the nervous system may process information in dysregulated ways, thereby giving rise to the full spectrum of psychopathological states and behaviors. It seeks to elucidate how psychiatric dysfunctions may mechanistically emerge and be classified, predicted, and clinically addressed" (ibid., p. 13).

In this endeavour, computational psychiatry came to merge insights and methods from the field of computational neuroscience – itself concerned with "formaliz[ing] the biological structures and mechanism of the nervous system in terms of information processing" (Seriès, 2020, p. 10) in terms of mathematical models – with recent changes in approaches to research in psychopathology, especially the research domain criteria (RDoC) (Cuthbert and Insel, 2013). RDoC is a research framework attempting to move beyond the supposedly stagnating current approach to psychopathology and treatment, by substituting the focus on psychiatric syndromes with a focus on mechanisms of specific dysregulations of cognition and behaviour relevant in the context of psychopathology. This approach was supposed to be better suited to integrating into psychiatry the increasing amount of knowledge gained from research on neural systems and behaviour in clinical and non-clinical populations. With this focus, RDoC and the attempt to use computational neuroscience for the purpose of psychiatric research have immense synergies, making them natural partners. As Seriès (2020) puts it:

> Rather than considering psychiatric diagnosis a cluster of symptoms, RDoC functional domains and constructs can be conceptualized as resulting from sets of underlying computations taking place across interacting neural circuits. In theory, these neural processes can, in turn, be described by algorithmic representations that describe information processing in the system. (p. 9)

10 Other earlier bird's-eye-view discussions of computational psychiatry can be found in, e.g., Montague et al. (2012), Walter (2013), and Friston et al. (2014).

Hence these neural processes could be described in terms of computational models, as used in computational neuroscience. Questions that research may at least in principle be able to address by pursuing these pathways would be questions such as "*What* are the main biological components involved in psychopathology and what are the mathematical relationships between these?", "*How* do dysfunctions in the individual biological units or in their interactions lead to the behavioral changes seen in mental illness?", and "*Why* have these changes occurred?" (ibid., p. 13).

Within computational psychiatry, we can differentiate between two broad classes of computational modelling: data-driven and theory-driven (Huys, Maia, and Frank, 2016). In data-driven modelling, machine learning is applied to large, multidimensional datasets from psychiatric patients, including genetic, neuroimaging, behavioural, and self-report data, and without considering any preestablished psychological or biological theories. Instead, the algorithm is supposed to find novel associations within the data structure that might give rise to new theories. Theory-driven approaches, on the other hand, attempt to provide a mathematical description of relations between types of behavioural performance or self-reports of psychiatric subjects and the performance of relevant biological mechanisms (such as brain anatomy or physiology) or higher-level functions (such as perception and learning) assumed to be relevant based on what we already know from previous work in computational neuroscience. By comparing the performance in self-report and behaviour with the underlying biological mechanisms and cognitive functions in healthy and clinical populations we may then generate a computation model of the dysregulations occurring in the clinical population.

Among the many examples of how computational psychiatry may in the future impact clinical diagnostics, I will select one from the branch of theory-driven computational psychiatry, and via this route return to my previous example, anhedonia. Anhedonia has more recently become an object of investigation in computational psychiatry (Kuhlmann, Walter, and Schläpfer, 2013; Huyes et al., 2013; Lambert et al., 2018; De Fruyt, Sabbe, and Demyttenaere, 2020; Walter, Wellan, and Daniels, 2020; Walter, Daniels, and Wellan, 2021; Liang et al., 2022).

Insights from research on reinforcement learning, including its neurobiological basis[11] and its relation to the phenomenon of pleasure, are especially important

11 Reinforcement learning is a strand emerging from the combination of two longstanding areas of theory: control theory and learning theory (Dayan, 2002). Control theory is an area of mathematics in which one attempts to provide value functions and dynamic programs that achieve optimal control of a dynamical system's behaviour. For this purpose, the theory attempts to identify a suitable control law for a system such that a given optimality criterion is matched by the system if the system is manipulated accordingly (Sutton and Barto, 2018). Learning theory, on the other hand, focuses on learning from trial and error and originated in psychology and the early investigations of animal learning in terms of Pavlovian (classical) and instrumental (operant) conditioning (Resorla, 1988; Staddon and Cerutti, 2003).

for this research. In the context of research on pleasure and its disruptions, phenomena are often considered in terms of the classical so-called pleasure cycle (Sherrington, 1906; Craig, 1918) assuming an appetitive phase (wanting), signified by the motivation for or the incentive salience of a reward; a consummatory phase (liking), signified by the pleasure of an actually achieved reward; and a satiety phase (learning) signified by representations and predictions about future rewards based on past experience (De Fruyt, Sabbe, and Demyttenaere, 2020). This basic model has been further developed by Rizvi et al. (2016), who describe the reward process as initially building a stimulus–reward association, which then leads to interest (wanting a reward), anticipation (a state of readiness for a reward), motivation (initial energy expenditure to attain a reward), effort (sustained energy expenditure to attain reward), hedonic response (enjoyment of reward), and feedback integration (updating reward presence and values). These aspects map quite well onto the aspects of the RDoC construct of positive valence systems: reward valuation (reward, delay, effort), reward responsiveness (reward anticipation, initial response to reward, reward satiation), and learning (probabilistic and reinforcement learning, reward prediction error, habit) (NIMH, 2018). On the neurobiological level, several regions are relevant, especially in the mesolimbic reward system consisting of a network of parts of the ventral tegmentum, the nucleus accumbens (part of the ventral striatum), and the amygdala (Schultz, 2002).[12] These regions are connected by dopaminergic signalling that seems to play a major role in reward-directed and consummatory behaviours in rodents as well as humans in general (Berridge and Robinson, 1998; Schultz, 2002; Egerton et al., 2009).

In however fine-grained a way we decide to think about anhedonia – whether we go with Rizvi and colleagues (2016) or with those researchers preferring a three-part model of wanting, liking, and learning (Bossini et al., 2020) – we end up with an understanding of anhedonia that, compared with that assumed in the DSM-5

Later evidence from lesion studies, pharmacological interventions, and imaging studies in animals and humans linked reinforcement learning with brain structures and functions of neurotransmitters, especially dopamine (Schultz, Dayan, and Montague, 1997; Heinz, 2017; Bogacz, 2020).

12 Besides these classically mentioned regions, other brain areas also appear to code and perhaps contribute to pleasure processing: for example, one site of the mid-anterior and mid-lateral part of the orbitofrontal cortex seems to track changes in subjectively reported pleasure (Kringelbach, 2005). For an overview of further regions and their (potential) implication in reward and pleasure processing, see Ellingsen, Leknes, and Kringelbach (2015). Due to the involvement of regions such as parts of the frontal lobe, researchers have proposed an alternative to the mesolimbic reward system in the form of the frontostriatal reward-processing network in frontal areas such as the ventromedial prefrontal cortex (vmPFC), orbitofrontal cortex (OFC), and midbrain limbic areas, including the ventral striatum (VS), insula, and thalamus (Sescousse et al., 2013).

discussed earlier (which assumes an impairment in wanting and liking), has more components, and therefore has more propositions whose presence might be evaluated as part of a propositional model to determine the presence of anhedonia. But since we are interested in how the improved understanding of anhedonia *qua* computational psychiatry might also impact the ways in which we diagnose, let us focus on this, instead of on the changes that we would see in a potentially new propositional model.

To return to diagnosis, let us look at studies that have used tasks to investigate the presence or absence of certain behavioural patterns and neural features in individuals suffering from anhedonia. Let us focus on research regarding the *wanting* component of anhedonia. Studies interested in this aspect have employed a variety of behavioural tasks, such as the "effort expenditure for rewards" task (Treadway et al., 2009), effort-based cost/benefit valuation tasks (Croxson et al., 2009), incentive motivation tasks (Anselme and Robinson, 2019), the "monetary incentive delay" task (Lutz and Widmer, 2014), reward-guessing tasks (Ubl et al., 2015), the wheel-of-fortune task (Dichter et al., 2009), and a slot-machine task for reward anticipation (Fryer et al., 2021).[13] While scientific evidence collected in these investigations is still not extensive, several interesting findings have been generated. I will focus on one of these. As a meta-analysis has shown, there are patterns of middle frontal gyrus and anterior cingulate cortex hyperactivation, as well as caudate hypoactivation, during different reward-anticipation tasks carried out with MDD patients, including monetary incentive delay tasks, card-guessing tasks, and wheel-of-fortune tasks (Zhang et al., 2013).

If we assumed for a moment that these findings are valid, in the sense that brain activation in individuals carrying out these tasks would show patterns of middle frontal gyrus and anterior cingulate cortex hyperactivation as well as caudate hypoactivation across these tasks if they suffered from the liking component of anhedonia, then these tasks combined with neuroimaging could be included in clinical diagnostic procedures to evaluate whether patients suffer from the symptom of anhedonia. The evaluation of this symptom would no longer be based on behavioural observations and self-reports of patients; instead, an objective bio-neuro-cognitive test could be used as part of the evaluation. Staying with this example, we may ask, would this step in the evaluation of anhedonia (or a similar step in this direction for any other psychiatric symptom) change the diagnostic procedure as described in my elaboration of my model-based account? The answer is: not necessary, but possibly.

Not necessarily, because the new psychopathological understanding of anhedonia can also be taken to offer the material for a different set of propositions telling us what it means for a patient to suffer from anhedonia and therefore for an alternative

13 For systematic overviews of behavioural tasks in combination with neuroimaging for the evaluation of reward processing, see Borsini et al. (2020) and Geugies et al. (2022).

constitutive propositional model of anhedonia. What may change given our neurobiological insights would then be an aspect of the assessment. After the screening phase of diagnostics that suggests a complaint that might be the psychiatric symptom of anhedonia, instead of evaluating this possibility by asking the patient questions or talking to their relatives, we might implement neuro-behavioural testing. If, for example, we took the proposition "Shows significant lack of motivation for initial energy expenditure to attain a reward (wanting component)" to be part of a propositional model of anhedonia, and we would accept that this lack is realised by a certain pattern of neural activity shown across monetary incentive delay tasks, card-guessing tasks, and wheel-of-fortune tasks. We might use these tasks and the recordings of brain activation patterns to evaluate the applicability of the proposition and thus the fit of this aspect of the model, via objective testing instead of self-report in the context of interviewing. Thus nothing changes in the overall order of diagnostic evaluation steps that I discussed in earlier chapters, and nothing about the use of model's changes. Only the means by which propositions are evaluated changes from interviewing to the new means of objective biological and cognitive testing – which, though so far for only a few psychiatric conditions, is sometimes already assumed to be part of the evaluation in the model-based approach.[14] In conclusion, it seems that changes that might occur as a result of developments in computational psychiatry could be readily integrated into the framework I have presented with my model-based account. However, when I said that our changing understanding of anhedonia would not necessarily change the procedure of diagnostics such that it would endanger the model-based account, I left open the option that it could do so. Let me come to this possibility now.

There are changes deriving from research in computational psychiatry – for example, in the research on anhedonia discussed here – that might in principle lead to changes in the overall diagnostic procedures in psychiatry that would make the account of psychiatric diagnostics discussed here obsolete. This would be the case if these changes impacted overall diagnostic practices and what is considered proper

14 For more examples of how computational psychiatry might inform diagnostics in a similar manner (i.e., by new means of evaluating diagnostic propositions), see Słowiński et al. (2017). They propose social biomarkers for identifying motor abnormalities that contribute to the deficits in nonverbal behaviours and in nonverbal synchrony that impair the structured and unstructured social interactions of schizophrenia patients, and that supposedly underlie patients' feelings of incompetence, confusion, and overwhelm in social contact, leading to the social withdraw of typical schizophrenia patients. The behavioural biomarker they use is motor behaviour in a "mirror game", a coordination task in which two partners are asked to mimic each other's hand movements, where the partner is a computer avatar or humanoid robot. With the help of statistical learning techniques applied to participants' movement data, they were able to provide a classification with 93% accuracy and 100% specificity.

diagnostic evaluation. Let us consider a few examples. If, for example, we developed neuro-cognitive objective tests for every single psychiatric symptom, it would in principle be possible to do no screening with patients as a method for deciding which potential psychiatric symptoms we should do an in-depth evaluation of. Instead, we might immediately have every patient do all the objective tests. We could move directly to the in-depth evaluation. While this might still be understood as an evaluation of the applicability of diagnostic models, this shift would change the procedure I discussed in the last chapter because there would no longer be a screening phase. As a result, the model-based account as it stands would be inadequate. Or take the current physiological and biochemical candidates for diagnostic biomarkers of major depressive disorder (e.g., Targum et al., 2022) or some of its symptoms (e.g., Stout et al., 2022) as measurable in clinical contexts. If they turned out to meet the specificity and sensitivity requirements for use in clinical contexts, they might supplement our current clinical practices. After identifying initial complaints that might indicate symptoms of depression, or that might point towards psychiatric symptoms that can occur in the context of major depressive disorder, we might then simply order the physiological or blood tests relevant to evaluate this possibility, providing us with a clear negative evaluation of whether the symptom or disorder in question is present. No mental modelling process, no comparing models to clinical observation, no evaluations of alternative sets of propositions that are part of qualitative models of symptoms would take place. Although there are still a number of problems in the pursuit of diagnostic biomarkers – such as underpowered and biased studies (Carvalho et al., 2020) for transdiagnostic biomarkers and low test-retest reliability and strong response to placebo intervention in psychophysiological biomarkers (Rapp et al., 2022), as well as ethical concerns (Glannon, 2022) – overcoming these obstacles and establishing biomarkers for clinical use would mean major progress in psychiatric diagnostics. If genuine, such progress would make my account a matter of philosophy of the *history* of psychiatry. These examples suffice to show the sensitivity of the model-based account to larger changes on the level of symptom diagnostics. Next, in order to underline the account's sensitivity to large-scale changes, let me come to changes in psychiatric diagnostics that are perhaps more unlikely to occur but are at least conceivable, and that might render the model-based account obsolete.

So far, I have focused on somewhat more realistic changes in psychopathology and clinical assessment that one might argue are already detectable in the current psychiatric literature. Now let me come to more extreme potential changes that would rapidly transform psychiatric diagnostics. These examples will make the point that in principle, such changes may falsify the model-based account. Let us consider two such scenarios. I will call the first one the Place-Feigl-Smart psychiatry scenario, the second one the Churchlandian psychiatry scenario.

What I call the Place-Feigl-Smart psychiatry (see Place, 1956; Feigl, 1958; Smart, 1959) would take place if two things were true. First, if the identity theory of mind and brain (i.e., types of mental states are identical to types of brain states) were correct, at least for those mental states interesting for psychiatry. Second, if we attain complete knowledge about how brain states and psychopathological mental states relate, such that these mental states and the behaviors they exhibit are fully intelligible in terms of structural or functional brain features. If this were the case, we would no longer need self-report, behavioral observation, or anything else from the patient. We would simply have to investigate their brain (let's say with some kind of neuroimaging) and let a program identify the present brain features that would then tell us what symptoms are present in the patient.

Alternatively, we may in principle end up with a Churchlandian psychiatry (Churchland, 1981) in which, since all talk of the mental in our language would be abandoned for brain talk anyway (to adopt Churchland's sketch of the future), pure brain and behavioural talk would also be all that we have when we talk about symptoms. Then mental symptoms would be out of the game and in their place we would have talk about brain states whose presence could be evaluated again by investigating the brain.

Although such radical scenarios seem unlikely to occur any time soon – even if the metaphysical framework that would have to be true to allow those scenarios to become reality were shown to be correct – what we can take from these two examples is that straightforwardly reading off symptoms from brain data would certainly make obsolete all the steps of the model-based account as spelled out here. When direct inference from brain data to psychopathological mental states which are mental symptoms or causes of pathological behaviour is possible, no modelling efforts as described by me seem necessary. We can also conclude that if we were, in a Churchlandian manner, to abandon mental talk entirely, the model-based account would collapse because we would drop talk about mental symptoms that need diagnosing from our diagnostic approaches altogether. Thus there would no longer be any efforts to engage in modelling to evaluate whether mental symptoms are present. The model-based account as presented would clearly be obsolete in both cases. Hence psychiatry could change in ways that would make the model based account an inadequate proposal to understand psychiatric diagnostic reasoning.

In conclusion, it appears that the model-based account is sensitive to changes in the reality of psychiatric diagnostics but at the same time general enough to encompass certain potential changes in psychiatric diagnostics. It is in touch with the reality of diagnostic practice and is thus a falsifiable theory of psychiatric diagnostic reasoning that is also not so overfitted that it loses all robustness against change. There is a spectrum of changes that it could integrate and accommodate.

4.8 Conclusion

In this chapter I discussed how the model-based account addresses the desiderata for a theory of psychiatric diagnostic reasoning providing an answer to the Methodological Question. I discussed whether the model-based account can be comprehensive and cognitively realistic, whether it helps us make sense of the difference between misdiagnosis and diagnostic malpractice, and whether it can account for the occurrence and resolution of diagnostic uncertainty, and concluded that it performs well in all these domains. Moreover, I argued that it helps us to understand and evaluate the phenomenon of good instinctual diagnostics and the occurrence and resolution of diagnostic disagreements. For each of these points, I set out how the model-based account fulfilled the criteria and thus meets all desiderata. In the next and final chapter, I will discuss alternative accounts to the whole of psychiatric diagnostic reasoning or aspects of it, and compare them to the model-based account to show the advantages it has over them.

6. Evaluating Alternative Views

In the previous chapters, I developed the model-based account of diagnostic psychiatric reasoning, arguing that it fulfils the adequacy conditions as well as the desiderata for an answer to the Methodological Question: "What is the method of proper, contemporary, psychiatric diagnostic reasoning?" Following on from the presentation of my own proposal, this last chapter will be dedicated to five other philosophical attempts to understand psychiatric diagnostic reasoning. The purpose of looking at alternative proposals in this chapter is to assess how my proposal holds up against them. If these proposals are compatible with my proposal, they may exceed mine in terms of convincingly meeting the adequacy conditions and fulfilling the desiderata; if they are incompatible with my proposal, they may present a more convincing proposal that also meets the adequacy conditions and fulfils the desiderata. We have just discussed the desiderata, so they should still be fresh in our minds, but let me offer a brief recap of the adequacy conditions.

The first adequacy condition was to adequately describe the method at work behind the diagnostic process. What does this method look like? What are its operations? When are which steps conducted? The second was to explain the rationale behind this method. What purpose do the steps of the method serve? How are these steps thought to contribute to the achievement of the epistemic ends of the methods used? The third was to set out how we should consider the justificatory status of beliefs achieved using this method. How are specific aspects of the method thought to justify its outcomes? Can we say something general about how promising the method is for arriving at true conclusions, or say how we may make such judgements for specific instances of the methods used?

With these conditions reviewed, for the purpose of comparing my proposal to the most relevant alternative views I will consider 1) Cooper's (2014) case formulation as an empathetic simulation account; 2) Murphy's (2012) sketch of diagnostic reasoning; 3) Reznek's (1988) inference-to-the-best-explanation account; 4) Gupta, Potter, and Goyer's (2019) intersubjective knowing account; and finally 5) Fuchs's (2010) and Parnas, Sass, and Zahavi's (2013) phenomenological approach to diagnostic reasoning. In my discussions of each of these views, I will show why my proposal is to be preferred as an answer to the Methodological Question.

5.1 Cooper: Individual Case Histories

In the fifth chapter of her book *Psychiatry and Philosophy of Science*, Cooper (2014) addresses the role of individual case histories as a means of explaining mental disorder in clinical contexts. Case histories as she discusses them should not be confused with the diagnostic output format of a case formulation. Case formulations as discussed throughout this thesis are primarily intended to provide an explanation for why a given syndromal diagnosis was chosen. In chapter 3, I interpreted the case formulation as a synthesis of information derived from the in-depth evaluation and the resulting selection of symptom models as candidates for the best explanation for a presented complaint of a patient – a selection that enables the choice of a syndromal diagnosis. Case histories as understood by Cooper, on the other hand, are a means to making a patient's behaviour and perhaps aspects of their cognition intelligible to us by looking at their life history. Case histories of patients, as Cooper puts it, provide "the beginning of an explanation of their behavior" (Cooper, 2014, p. 69). At first glance, one might, as has been suggested by some philosophers (e.g., Murphy, 2020), think that Cooper's case histories provide an alternative proposal to my understanding of case formulations. One may think that while my proposal intends to enable diagnosis by identifying symptoms through the use of constitutive models and hence takes a constitutive approach to explaining patients' dispositions to produce the occurring symptom, Cooper's proposal uses patients' life stories to provide a causal approach to identifying present symptoms. As I will argue below, however, interpreting Cooper's account as an alternative to mine is wrong. First, though, let us explore further what kind of explanation Cooper is aiming to provide with a case history.

The very purpose of case histories, as they are usually understood according to Cooper, is "a narrative understanding, empathy, or 'verstehen'" (2014, p. 79) regarding a patient's complaint. This understanding is provided by an "explanation of why they thought as they did is some particular circumstances" – an explanation that, although we are considering a specific case, "will be an explanation of why any human being would think in that way in that circumstance" (ibid, p. 70). Cooper's proposal is that what the clinician is doing when they try to achieve this simultaneously general but also specific understanding of a patient's psychology is to wonder what they themselves would have done.

This act of self-reflection is interpreted by Cooper in line with the simulation account of folk psychology that has been put forward in varying forms by several philosophers (e.g., Heal, 2003; Goldman, 2006; Hurley, 2008). Roughly speaking, the basic idea behind the simulation account of folk psychology is that we imagine (either unconsciously or with conscious effort) being in another person's position based on what we know about them and their situation and run a simulation of what we would do or think if we were them in order to understand their current or predict their future behaviour. Although there are many nuances to this process, this is also the level of abstraction that Cooper operates on when describing this account, so we will adopt a similar descriptive level here. In light of this background, she claims that case histories basically work in the same way:

> Case histories work by providing us with the scaffolding to simulate another. This explains why case histories focus on all that is unique to the individual. I can suppose that most of another's mental states and ways of thinking will be the same as my own (they too will think that $2 + 2 = 4$, that Paris is in France, that good food is nice, that being wet and cold is bad, and so on). As such, it is their peculiarities that I need to know about if I am to make necessary corrections to my own ways of thinking to be able to mimic theirs. Along similar lines, the more detail provided by a case history the better it will tend to be. The more information I am given about another, the easier it will become for me to think as if I were them. (Cooper, 2014, p. 69)

As an example, Cooper presents the sketch of a case history of Mary, a patient of the psychotherapist Robert Akeret (1995):

> Akeret's patient, Mary, had a Catholic upbringing. She had been brought up to believe that evil thoughts are approximately as bad as evil actions. As a child, on a number of occasions she had wished that bad things would happen to people, and they did. One day she became angry with her father and wished he were dead, and the next day he died. On the basis of this story, we can easily imagine how we would feel if we had Mary's beliefs and were in her situation. It will not come as a surprise to us that Mary suspects it is her fault that her father died, and that this leads to feelings of guilt and depression. (Cooper, 2014, p. 69)

Providing an explanation by means of simulation is, as Cooper readily admits, not a very deep kind of explanation. As she puts it:

> In so far as the target system can be simulated, the explanation of its behaviour must refer to features that are shared with the simulating system. Of course, we may still want an explanation of why it is that *any* of the systems behave as they do. When we simulate a system, this does not completely explain its behaviour, but it does at least tell us what kind of explanation we should look for. (ibid., p. 70)

More important for my purpose of discussing this account here, however, is an epistemic feature of a case histories used in this way. According to Cooper, mental states and behaviours of patients that can be accounted for in this way are not abnormal:

"insofar we can simulate them, we can conclude that there is nothing special or abnormal about the subject" (ibid.), she says.

Cooper adds to this that simulation might not be possible in all cases. I agree with her, and this will be important for my interpretation of her account as well as in later sections where I criticise other accounts that also rely heavily on empathy. But why does Cooper think so? Not all experiences or cognitive or emotional states, and therefore not all behaviours, are necessarily open to simulation. Someone who has never experienced hospitalisation may have a hard time understanding the behaviour of people seeking to avoid it. It might also be hard to understand the reactions to certain situations that are manifested by people who have experienced torture. Or, to consider the example of delusions, there might be cases in which understanding a patient is still within the realm of the imaginable – such as patients who have the delusion that spots on their face contain maggots, which we may imagine in terms of weird skin sensations causing us to want to get something out of our skin. Other delusions, however, especially concerning emotions and puerperal beliefs, might be harder to imagine – like the delusion of having a romantic relationship with the polar beer in the local zoo, as Cooper suggests (2014, p. 76). It might be even harder to imagine and therefore understand the thoughts and behaviours of someone suffering from Cotard's delusions, perhaps claiming to have rotten organs, not to have eaten or slept for years, or to have no blood and indeed be dead but still here (ibid., p. 77). So much for Cooper's account. Next, let me turn to the question of how her account relates to my proposal.

The first question to ask is whether her account is compatible with mine or not, and whether it covers any aspects of diagnostic reasoning that my account neglects. It may initially seem that there is tension between Cooper's account and mine, because one may perceive a contradiction between her proposal for how to understand the case history and my ideas about the nature and purpose of the case formulation. I do not think that this is the case.

Although the case history as well as the diagnostic case formulation draw on information about the patient's past experiences, behaviours, and social circumstances and employ them to explain something about the patient, they do so in different ways and for different purposes. Whereas the purpose of the diagnostic case formulation as part of the diagnostic proposal is to serve diagnostic classificatory diagnostic interest, the purpose of the case history is not classificatory but to enable a narrative (folk-psychological) understanding of what the patient is doing and experiencing. The case history is therefore trying to do something different from a diagnostic case formulation. The case formulation serves the aim of backing up the classificatory decisions that are ultimately expressed in a syndromal diagnosis, in my opinion by summing up the decisive evidence that led to choices for and against symptom models. The case history, on the other hand, allows the clinician to understand aspects of the patient's experiences and behaviour in an empathic way, which might be useful for building a working relationship with the patient because the patient feels understood by the clinician, or to acquire a sense of what therapeutic strategies might be employed to help the patient. For example, if there is a plausible folk-psychological understanding of some of the patient's problems, there might be some obvious way to help – such as reducing stress reported by the patient who reports being totally stressed out in a way that is quite understandable from their situation. However, this relationship-building and potentially interventional value is not the same as diagnostics.

That case histories as considered by Cooper are not meant to play a central diagnostic role in psychiatry and therefore should not be considered a potential alternative to diagnostic case formulation. My way of understanding case histories becomes clear when we consider the limitations that Cooper herself points out. Cooper suggests that there are many non-typical mental states that are perhaps hard to grasp for a clinician *qua* mental simulation based on a case history. As one example she proposes specific delusions such as the delusion of having a romantic relationship with the polar beer in the local zoo. Given their uncommonness, we can perhaps assume that there are other psychopathological phenomena that are difficult for clinicians to simulate, such as the experiences of people who are so severely depressed that they show mutistic behaviour, stop eating, and stop getting out of bed. Another example that is perhaps hard to imagine for someone who has never experienced it would be a full-blown panic attack. If case history-based simulations were the method of diagnostics, we could perhaps not diagnose delusions diverging so far from common experience as well as other psychopathological conditions as for example panic attacks, or depressed mood seen in especially severe cases of depression, as we had a hard time simulate them. Since we do diagnose these disorders, and since in these diagnostic processes (as in any diagnosis) a formulation is expected to do the explanatory work for the resulting diagnosis, it seems that case histories cannot be an alternative approach to case formulations – at least unless Cooper expressed some scepticism towards diagnosing such empathically challenging conditions, which she does not. So, if Cooper's account is apparently not trying to provide a theory of diagnostic case formulation under another label, does her approach – and folk psychology along with it – really have no relationship with diagnostics? Not even in part? One might think this strange. Indeed, folk psychology plays a role in psychiatric diagnostics and the case formulation.

According to my own approach spelled out in Chapter 3, folk psychology plays into the process of model-based psychiatric diagnostics in the evaluation of psychological complaints. As I discussed there, complaints may be evaluated *inter alia* as non-pathological psychological problems. In this case, they are not classified as symptoms of a psychiatric or other medical disorder. This outcome will be reached if a propositional model supporting this no-symptom evaluation is best (and sufficiently well) supported by the diagnostic information about the patient – in other words, if the occurrence of psychological complaint is constituted by circumstances that are judged to render it a normal mental occurrence. What renders a psycho-behavioural reaction normal rather than pathological, and as such occurs as a proposition in the model supporting this judgement, will be influenced by the understanding of normal psychology employed by the clinician. This understanding will in turn be influenced by psychometric knowledge and academic psychological knowledge about normal psychology, and also by folk psychology.

Take the example provided by Cooper: Mary, who wished that her father would die before he died in an accident and who believes that evil thoughts are as morally wrong as evil actions. If her father died yesterday and she reports such feelings no clinician would judge her guilty feelings to be a psychiatric symptom; rather, they would appear to be an immediate psychological reaction in line with her moral convictions. In this context, her guilt does not appear to be pathological, it is not (for example) a delusion, and since it is acute and guided by moral conviction it does not seem to be rumination. Her presentation is constituted by factors that would lead to the evaluation of non-pathologically relevant psychological distress. As this example shows, folk psychology can and will often have a place in psychiatric diagnostics, namely as a background theory based on which propositional models of psychological complaints that would render them non-pathological can be set up. Folk psychology and its uses for understanding others, however, are not the whole engine of psychiatric diagnostics.

I conclude this section by summing up some core points discovered in the discussion of Cooper's work. Although at first glance it might seem as though Cooper's proposal and mine are competing to explain how information about patients is used to provide an overarching representation of their case for the purpose of drawing diagnostic conclusions, this is not the case. I demonstrated why Cooper's case histories are different in nature and aim from the model-based account of case formulations: While case formulations aim primarily to support and back up diagnostic classification, case histories support the relationship-building and interpersonal understanding between patient and psychiatrist on a folk psychological level. This can be useful for several clinical purposes, but it is not intended or equipped to be a tool for proper clinical psychiatric diagnostics.

5.2 Murphy: A Version of Diagnostic Modelling

In his book *Psychiatry in the Scientific Image*, Murphy (2012) addresses the issue of psychiatric diagnostics. He provides a very brief discussion of his idea of diagnostics in psychiatry, which even makes reference to some of the same literature on philosophy of modelling that I discussed in previous chapters. But although Murphy talks about modelling in the context of psychiatric diagnostics, I will show that his account and mine are vastly different. It is not straightforward to decide whether his proposal should be understood as aiming to provide a full understanding of how psychiatric diagnostics works. But no matter how one reads his proposal, be it as one that claims to provide an account of psychiatric diagnostic reasoning in general or only in some of its aspects, I would argue that his account is insufficient. It would be insufficient as an overall proposal for answering the Methodological Question because Murphy does not address all adequacy conditions and does not meet relevant desiderata. If, alternatively, we interpret his account as a proposal for only some aspects of what would be needed for a complete answer to the Methodological Question, his account would also be insufficient. In this case, it is insufficient because even the aspects of the Methodological Question that he does address – which, as we will see, are the descriptive adequacy and the justificatory adequacy condition – are addressed in an implausible manner. But before I come to argue all this, let us begin by looking at his proposal.

Murphy's approach starts from the assumption that psychiatric disorders are usually thought of as exemplars, by which he means "idealized theoretical representations of a disorder" (2012, p. 206), and that they must be differentiated from models. Models, according to Murphy, go beyond exemplars:

> An exemplar is a representation of the typical course and symptoms of a mental illness, whereas a model is a representation of those symptoms, that course, and the causal determinants of both of them. A model is an exemplar together with an explanation. (ibid., p. 206)

He also puts it slightly differently, with more emphasis on the nature of what he means by causal determinants:

> [A] model is an explained exemplar: the exemplar is the typical manifestation of the symptoms and course of disorder, and a model is the representation of the causal relations that obtain between features of the exemplar and various aspect of the organism. (ibid., p. 207)

Murphy goes on to explain his take on diagnostics considering this understanding of an exemplar of a disease:

> diagnosis works by fitting a patient to a portion of the exemplar, and the exemplar is explained by modelling the process whereby the symptoms in the exemplar express the state of neurobiological system (pathology) that depend in its turn on logically prior causal processes (etiology). (ibid., p. 206).

Murphy's brief remarks are more a sketch than a full-fledged proposal of how psychiatric diagnostics is supposed to work, but they provide enough detail to be evaluated and compared to my proposal.

A core difference between Murphy's account and mine concerns our perspectives on the role that models and modelling play in diagnostics. While my view is that models are set up based on background knowledge to be used to diagnose symptoms in a process of comparing these models to the patient, Murphy assumes modelling to play a vastly different role. In Murphy's account there are no models of specific symptoms, as I propose, but only models of disorders. Moreover, while models play a direct role in the diagnostic evaluation of the patient, in his account they are only the background from which features to look for in patients are derived. So, how should we assess Murphy's account?

I will argue that Murphy's approach has two problems. First, it does not meet the adequacy conditions for an answer to the Methodological Question. And second, the proposal he makes does not insufficiently address some of the desiderata of an answer to the Methodological Question. The proposal does not meet the adequacy conditions because among these condition (providing a description of the method at work at psychiatric diagnostics, providing a rationale for the inferential processes at work within the proposed methods, and providing an understanding of how the outcomes are supposed to be considered justified), he at least fails to meet the descriptive criterion, the rationalisation-of-inference criterion, and at least to some degree also the justification-related criterion. Moreover, his proposal does not enable us to address several desiderata in a sufficient manner, or at least does so in a less satisfying manner than the model-based proposal does.

That Murphy's proposal does not provide a rationale for the inferential operations undertaken in the process of diagnostics. What kinds of inference are made and how the inferential patterns employed are supposed to support his conclusions is essentially not discussed by Murphy. He tells us that the exemplars of disorders are compared to the patients to decide the outcome. However, what kind of inference is taking place and how exactly any specific type of input is enabling the inferential matching to work to produce its outputs is not addressed in any detail. It thus seems fair to say that this adequacy condition is simply not addressed by Murphy's proposal. Next we turn to the adequacy condition of illuminating why we should deem the outcomes of the method's inferential work justified.

Murphy makes no proposal regarding internal justification; he cannot, because he has not spelled out the structure of the inferential method he proposes sufficiently well to make claims about how it is supposed to provide justification. However, his claims about where the exemplars come from that are used in diagnostics might be considered as a proposal for where the external justification is coming from: namely, the scientific models used to set up the diagnostic exemplar. In principle, this seems reasonable. After all, when I talked about external justification in the last chapter, all I did was gesture towards the science of psychopathology. However, there is a problem with Murphy's attempt to rely on science to obtain justification for his diagnostic proposal. He assumes that the psychiatric sciences add something to the process of diagnostics that it cannot offer. Thus he has not provided an acceptable approach to the external justification of his proposal. Let me elaborate.

Murphy assumes a support for the diagnostic exemplars that is problematic because he seems to have an inadequate picture of the state of psychiatric knowledge and its application to psychiatric diagnostics. Murphy's approach seems to presuppose that there are widely accepted explanatory models of psychiatric disorders informing us about the proximal causes (i.e., physiological processes) giving rise to certain symptoms as well as about the distal causes that brought about the changes responsible for the presence of the psychiatric symptom. This is what he assumes models in psychiatry to present us with. The exemplars then used in diagnostics are basically this model minus the explanations; they contain only information about the symptoms explained by the model, as well as the cause of their occurrence and change in the context of the disorder. This is a highly problematic background assumption. As I mentioned in Chapter 3, despite many interesting and important scientific efforts, psychiatry currently lacks full-fledged detailed models of psychiatric disorders as a whole – and even for most psychiatric symptoms – that would offer a detailed mechanistic explanation of the proximal and distal biological causes of occurring symptoms, as well as of the developmental pathological importance of various factors such as genes and social environment.

Moreover, beyond the face-value fact that there are no such models around yet, it even seems implausible that there could be anything like such a unitary model for many major psychiatric disorders according to the currently used diagnostic classifications, because many disorder are likely lumping together clusters of distinct conditions. Just think of major depression. Major depressive disorder can occur in patients with 227 combination of symptoms that are vastly different and partly without any symptomatic overlap, which, according to our best current scientific understanding, suggests that vastly different causal (e.g., neural) processes are involved in different instances of one and the same disorder (as classified in current diagnostics). This is all the more likely if we consider instances with no symptomatic overlap, which we know are not only possible according to the manual but indeed occur in significant numbers in patients (Zimmermann et al., 2015). If we assume that different symptoms and especially non-overlapping or only partly overlapping clusters of symptoms will be caused by non-identical psycho-biological processes, there cannot be one scientific model of major depression, because major depression is not a single phenomenon but seem to consist of multiple phenomena that science would have to identify and explain. One model could not comprehensively cover everything that falls under the label of major depression. Accordingly, even if we had good causal models of psychiatric disorders, the case of major depression illustrates that, given the current diagnostic systems, we would perhaps not end up with one only. Hence, there could not be an exemplar whose components could then be fitted to the patient. In conclusion, it therefore seems that Murphy's view of the state of psychiatric diagnostics and the way it can enable clinical diagnostic work fails to make a adequate proposal for an answer to the Methodological Questions that is true to the state of psychiatric science and diagnostics.

Considering the problem with the assumed unitary background models of mental disorders that, according to Murphy, is meant to back up and justify the exemplars used in psychiatric diagnostics, his proposed method of psychiatric diagnostic seems implausible regarding the external justification condition. Internal justification is not addressed by him at all. Hence, the adequacy condition of spelling out how conclusions of the used method are supposed to be deemed justified seems to be failed by Murphy's proposal. What about the descriptive adequacy condition, to propose a method via which psychiatrists draw diagnostic conclusions that maps onto the diagnostic efforts of clinical psychiatrists? After all, one could say that the idea of comparing exemplars to patients seems to provide such a proposal and that it is not so far removed from my position that disorder diagnostics takes place as pattern recognition. This seems to be a plausible proposal for a method, and even one where we seem to agree with each other, but Murphy made this point before me. I disagree, or at least I would claim that interpreting my way of describing the intermediate steps of drawing diagnostic conclusions as just another way of putting what Murphy had in mind would be as unnuanced as the worn-out claim that Plato already said everything there is to be said in philosophy, However, this depends on how exactly we understand Murphy's proposal. Let's look at it again.

In his proposal, describing the belief-forming procedure – that is, the method by which psychiatrists arrive at diagnostic conclusions – he claims that exemplars, which consist in assumptions about sets of symptoms and the course of their development derived from a background model of the disorder, are used in a process of "fitting a patient to a portion of the exemplar". Let's accept this idea and forget for a moment that, as I argued earlier, such exemplars cannot be derived in the way Murphy proposes, instead focusing on his proposed method, the "fitting [of] a patient to a portion of the exemplar". It appears that there are at least two ways to understand this short phrase and therefore the proposed processes of diagnostic reasoning according to Murphy: one that appears to be highly problematic and should for reasons of charity not be attributed to him, as this would render his proposal a failure, and one that is indeed more plausible and closer to my own ideas, but so underdeveloped and implicit in his writing that one could hardly argue that Murphy made the same proposal as I did, given that developing the proposal to an adequate level of detail is part of the heavy lifting I undertook in the last chapter. By either of the readings, it would seem that Murphy's proposal to describe the method either fails to be adequate or is at least less adequate (because it is not worked out in any detail) compared to mine. Let's begin with the more problematic reading of what he may mean by "fitting a patient to a portion of the exemplar".

One way to understand "fitting a patient to a portion of the exemplar" would be to assume that psychiatrists somehow evaluate patients for the presence of fixed packages of symptoms making up whole disorders that would be sufficient to provide a psychiatric diagnosis and that this is the one and only level of diagnostic evaluation. However, if we accept this, there would be no lower-level diagnostic investigation as part of the diagnostic evaluation – that is, no inferential process that evaluates the patient for the presence of specific symptoms so that patterns of symptoms required for a diagnosis can be identified in the output of such a lower-level diagnostic process. If this were what Murphy wanted to say, his approach would seem implausible. On the one hand, it would ignore all the diagnostic reasoning work of the psychiatrist that contributes to deciding whether a symptom is present or not. Moreover, it seems that there are diagnostic categorisations whose assignment to a patient could not be carried out by Murphy's approach. Think, for example, of the categories of "unspecified depressive disorder" (APA, 2013, p. 184) that allow a psychiatrist to diagnose a depressive disorder if several psychopathological symptoms of depression are present, but not all necessary criteria for another depression diagnosis are fulfilled. There is no concrete description of the exact number of combinations of depressive symptoms that need to be present for this disorder to be diagnosed. It seems hard to imagine that Murphy wants to claim that there is an exemplar that represents all depressive presentations that do not fulfil any other depression-related condition requirements, given that an exemplar, according to Murphy, is an "abstract" and "ideal" representation of the disorder. It therefore seems that a diagnosis is intended to be provided based on previous insight into the presence of psychiatric symptoms and recognition of one of many potential patterns of symptoms that do not suffice for any other depressive disorder diagnosis and thus yield this diagnosis. However, this requires a diagnostic reasoning process that identifies symptoms in the first place, which is not part of Murphy's proposal as interpreted here. Another similar point about Murphy's proposal is that it would not explain how the psychiatrist may recognise symptoms insufficient to support any disorder diagnosis but occurring somewhat disparately and not feeding into any of the disorder diagnoses given to the patient. A patient might, for example, suffer from minor depression but also experience depersonalisation. How could the psychiatrist be aware of this single symptom if it were not acknowledged by the application of a disorder exemplar? It seems again that some lower-level diagnostic reasoning process is necessary for this that goes beyond the application of exemplars to patients. However, there is a different, perhaps more plausible, and realistic way to interpreting "fitting a patient to a portion of the exemplar".

On this second interpretation of "fitting a patient to a portion of the exemplar", we could take Murphy's account to imply that the psychiatrist knows what a disorder would look like if all its symptoms where present and knows what combination of subsets of these symptoms, which Murphy would call "portions of the exemplar", would need to be present in order to provide the diagnosis. If we understand Murphy along these lines, his idea would indeed be compatible with my proposal, as this is basically what I also assume that psychiatrists are doing. Murphy, however, taking the first steps on the path I have taken with my proposal, did not flesh out this idea to any grain of detail comparable to the proposal I have made in the preceding chapters. Accordingly, even if we understand Murphy along these lines, it seems that my proposal exceeds his in detail and explanatory depth by a wide margin, so that again it appears fair to say that Murphy's proposal does not adequately explain in detail what goes on in the process of psychiatric reasoning, even if we are willing to grant that he intended to imply what my proposal worked out explicitly. Thus, once more it seems that his account lacks the criterion of providing a description of the method adequate to diagnostic practice because it fails to address relevant aspects (symptom diagnostics) in detail. We may assume, in this more charitable interpretation, that his account implies a more detailed explanation, but he does not say how symptom diagnostics is supposed to take place. The lack of detail in Murphy's account of symptom diagnostics, and his rather abstract way of talking about the disorder diagnostic part of the proposal, can on the most generous reading be understood to fulfil the descriptive criterion for an answer to the Methodological Question to a small degree, and certainly to a lesser degree than my proposal, which also details the steps of the method of diagnostics on the symptom level. This makes his proposal a weak substitute for mine.

If we sum up by asking how Murphy's ideas hold up against the three adequacy conditions for an answer to the Methodological Question, it seems that he scores low. The criteria related to justification and the rationale for inferential patterns were not provided or were shown to be implausible. The method description was present in an insufficient manner on the most charitable interpretation. It therefore seems that Murphy's ideas represent an inadequate attempt to understand the method of proper contemporary diagnostic reasoning. Although it is no longer needed because the proposal is already shown to be inadequate, let us nonetheless talk briefly about desiderata. What Murphy presents us with would seem also to fail many of the desiderata. His account is certainly not comprehensive, since it fails to talk about the whole aspect of diagnostics in enough detail to understand what happens there (symptom diagnostics) and leaves out whole aspects of clinical diagnostics (i.e., diagnostic co-formulations resulting from critical discussions between clinicians). Moreover, those aspects of diagnostics that are addressed in his proposal are explained in such an abstract way that they hardly seem to have the explanatory resources to provide a remotely detailed understanding of, for example, the difference between misdiagnosis and diagnostic malpractice, diagnostic disagreements and their resolutions, diagnostic uncertainty and how to resolve it, or how good diagnostic instincts may work.

In sum, Murphy's proposal fails to meet any of the adequacy conditions for an answer to the Methodological Question, or at best meets one of them to a very limited degree. As mentioned at the beginning of this section, this means that it fails as an alternative to my model-based proposal, no matter whether we take his proposal as a full-blown attempt to address the Methodological Question or only as some ideas addressing just a subset of its central requirements. As briefly discussed at the start, it also seems that there are at least several desiderata for an answer to the Methodological Question that Murphy's account seems unable to fulfil. As I have demonstrated in the preceding chapters, the model-based proposal, by contrast, meets all the conditions and is able to fulfil the desiderata, so it seems fair to conclude that the model-based account is to be preferred over Murphy's ideas about psychiatric diagnostics.

5.3 Reznek: Inference to the Best Explanation

In his article "On the epistemology of mental illness", Reznek (1998) discusses the challenges of psychiatric diagnostics and puts forward a proposal for how psychiatrists arrive at justified conclusions about the presence of mental disorders or psychiatric symptoms in patients. As such, Reznek's proposal should perhaps be understood not as an attempt to provide a full answer to the Methodological Question, but rather as an effort to address two aspects of an adequate answer to it: what patterns of inferences are at work in psychiatric diagnostics and how its conclusions using these patterns of inference may be deemed justified. As I discuss below, Reznek's ideas about how to address these two aspects overlap to some extent with mine, but my position offers a more satisfying answer to these two aspects of the question. Furthermore, by addressing the remaining aspects of an adequate proposal and also fulfilling the desiderata of an answer to the Methodological Question, the model-based description of the psychiatric method proves preferable – regarding the specific aspects of diagnostics that both proposals addresses and also as an overall more satisfying framework. Let us begin by looking at Reznek's framework.

Reznek's starting point is the well-known Rosenhan experiments (Rosenhan, 1973). In a nutshell, Rosenhan sent supposedly mentally healthy people to psychiatric hospitals, instructing them to pretend to hear voices. These individuals were diagnosed with psychosis and admitted to treatment. Reznek treats this occurrence as a case study bringing to our attention a problem for psychiatric diagnostics that he calls the "Rosenhan challenge":[1] "there is a logical gap between the description of subjective symptoms and the attribution of an objective disorder. We cannot deduce the presence of a disorder from a list of purely subjective symptoms or behaviours" (1998, p. 216).

After discussing alternative approaches to how one might arrive at conclusions about the mental states of others and how they might enable psychiatric diagnostics, which, according to Reznek, fail to provide sufficient support to diagnostic conclusions, he arrives at the only approach he considers promising. He calls it the scientific or hypothetico-deductive method. This method "postulates the existence of some theoretical entities to explain observable phenomena" (1998, p. 218). To map this method and its scientific use onto attempts to determine the presence or absence of a mental state in other, he presents an illustration:

> For example, when Newton observed such diverse phenomena as the tides, the motion of the planets, and the falling of apples, he hypothesized the existence of the gravitational force that explained such observations (even though Newton claimed that he never made hypotheses). We come to believe there is such thing as a gravitational force because we need such a theoretical entity to explain these observations. In the same way, we might postulate the existence of mental events – they are the theoretical entities that are needed to explain behavior of other people. Without them we cannot make sense of their behaviour. The explanatory power of such theoretical entities provides evidence for their existence (just as it does in science). This seems our most reasonable approach. (ibid.)

Interestingly, what is being described here by Reznek is not the hypothetico-deductive method as usually conceived since Popper (1935), but rather an abduction or inference to the best explanation. Let me elaborate.

The hypothetico-deductive method usually follows an algorithm that contains more elements than the one described by Reznek. Let's take the explication offered by Godfrey-Smith (2003, p. 236). According to Godfrey-Smith, the hypothetico-deductive method proceeds roughly as follows: (i) Use of experience: You consider a problem/observation you wish to explain and gather data about it. (ii) Forming a conjecture: You put forward a hypothesis whose truth would adequately explain the phenomenon of interest and the data you gathered about it. (iii) Deducing predictions: You deduce predictions that must follow from the truth of (ii). (iv) Testing: You

1 It may be worth emphasising that Reznek himself does not buy the sceptical conclusions that Rosenhan himself drew from his experiments regarding the validity of psychiatric diagnostics, but merely considers Rosenhan's work to put forward an interesting challenge. Earlier responses to Rosenhan's work challenged the power of Rosenhan's experiment to support his sceptical conclusions altogether (Spitzer, 1975), and more recent responses have presented evidence of massive fabrication of data in his studies (Cahalan, 2019).

consider observational evidence that could disprove the predictions and therefore refute our hypothesis. As is widely known, however, in a purely hypothetico-deductive framework, evidence not conflicting with the hypothesis is usually not considered evidence for the hypothesis, as this would equal the logical fallacy of affirming the consequence.

Steps (iii) and (vi) are missing from Reznek's example, where it appears that the explanation that is considered to make most sense of the observations is what is accepted as an explanation. This suggests that what is happening here matches with the pattern of inference to the best explanation, which, according to Lipton (2017), has as "[i]ts governing idea is that explanatory considerations are a guide to inference, that scientists infer from the available evidence to the hypothesis which would, if correct, best explain that evidence" (Lipton, 2017, p. 184).

One may consider this a misreading of Reznek. Maybe he intended his example to contain steps (iii) and (vi) but he did not try to explicate them because he expected his readers to be sufficiently primed by his mention of the hypothetico-deductive method to do this for themselves. Textual evidence speaks against this reading.

Later, Reznek gives another example of his preferred account to diagnostics, discussing how to diagnose whether someone is suffering from hallucinations:

> We identify genuine hallucinations by comparing two overall hypotheses of bizarre behaviour – one is that the person is hallucinating (and is deluded), and the other is that the person is malingering. The hypothesis that provides the best overall explanation of the behaviour, and is consistent with all our knowledge of ethnology, anthropology, and so on, is the one we should accept. We will have no proof, but only a good hypothesis. But since this is all we have anyway, in any discipline, we should not feel uncomfortable. (1998, p. 229)

Again, this paragraph highlights that what the psychiatrist is doing is a comparative judgement amongst different explanations, choosing the one that is supposedly most coherent with other theoretical assumptions as well as with the observations at hand rather than facing a test in other situations – that is to say, no step (iii) or (iv). Moreover, what is even more striking is that in this example it becomes even clearer that in the context of diagnostics, the generated hypothesis explains not only potential future phenomena but also the specific phenomenon at hand that led to the psychiatrist formulating and choosing among the explanatory hypotheses. This again seem to be a feature of inferences to the best explanation rather than of hypothetic deductive reasoning. As Lipton (1991, p. 67) points out, one advantage of inference to the best explanation over the hypothetico-deductive method is that while hypothetico-deductive explanatory entitlement is directed *only* to future events (i.e. the hypothesis generated in (ii) is only applied in step (iii), not the very phenomenon that inspires the hypothesis to be formulated) Inference to the best explanation has a broader scope of explanation. Namely that "Inference to the Best Explanation suggests that explanatory considerations may apply to both the generating candidates and the selection from among them" (ibid.). In other words, in contrast to the hypothetico-deductive method, Inference to the Best Explanation not only explains future events but also the context in which it was initially conceived.

In sum, it appears that Reznek is confusing abduction with hypothetic deduction. Accordingly, his answer to the Rosenhan challenge is that although we cannot deduce mental symptoms of patients from their behaviour, we can make inferences about their presence by taking into account everything relevant we have learned about the mental phenomena we are considering attributing to the patient and assessing whether the patient's presentation makes it seem most coherent that this phenomenon is occurring in the patient or whether another explanation is more plausible. We can do so by an inference to the best explanation, which would involve the assumption of the presence of the condition in question.

How do we learn about the phenomenon in the first place if we cannot determine its presence directly through the observation of behaviour? Reznek proposes that we must start with some stipulations to carve out a phenomenon, which on his view is a matter of clinical judgement:

> We are first required to decide who is depressed, for example, and who is not. Only after this, can we find out what sorts of questions best identify those who are depressed. These questions can only be as good as the clinical skills that differentiated the two groups in the first place. And the test can only be as objective as the diagnostic process that set up those groups prior to the construction of the test itself. Far from being an objective test of psychiatric disorder, the psychological tests are as subjective as the clinical procedure on which they are based. (1998, p. 223)

Ideally, Reznek goes on, such decisions should be made according to what he takes to be the ideal case in medicine – namely, with reference to biological disease underlying an occurring disorder that is described in terms of a symptom-based syndrome. As he puts it, "In medicine, the identity of a disorder is defined by the underlying biological nature of the syndrome rather than the syndrome itself" (ibid., p. 219).

Assessing Reznek's proposals, there are several synergies between his account and mine. We both believe that psychiatric diagnostics is not a straightforward deductive inference from utterances or simple behaviours of patients to the attribution of a symptom, precisely because the simple occurrence of a behaviour underdetermines what is going on with the patient. We both assume that as part of the assessment we make an inference to the best explanation to what condition is present in the patient on the symptomatic level, given the relevant evidence we have collected about the patient. And we both believe that our psychopathological understanding based on background knowledge about these psychopathological conditions is central to this inference.

I have no bone to pick with Reznek about the things he says. In terms of internal and external justification of psychiatric diagnostics, he tells us that psychiatrists' inferences are inferences to the best explanation about which conditions' presence should be assumed as the best explanation of the patient's presentation. He thus presents us with the inferential pattern that rationalises the diagnostic process, and he indicates how diagnostic judgements are supposed to be internally justified. He then also tells us where the credibility of judgements made this way comes from – namely, from the scientific understanding of the psychiatric condition via which we calibrated our initial judgements about what should be considered the explanation-worthy phenomenon in the first place. This provides a proposal for external justification. Although I agree with all this, I think however that the model-based account has more to offer than Reznek's, including if we look at the very topics also addressed by Reznek.

On the point of meeting the Methodological Question's requirement to provide a rationale for the method used in the diagnostic process, Reznek can only say that whatever precisely the method is (he is not proposing a concrete description of a method), it works *qua* the inferential pattern of inference to the best explanation, and then gives us some examples. This may be right, but the lack of a proposed method makes this answer rather abstract, since there are many ways in which an inference to the best explanation can take place. Reznek describes, in his examples, how information about the patient is collected and taken to point towards a diagnosis based on our understanding of what kinds of behaviours and experiences we should expect to see in a patient if he has this diagnosis. My sense is that all this tends in the same direction that I have pursued in my more detailed proposal – namely, that the diagnostic process follows an indicator (in my argument a constitutive indicator) strategy. This idea may lurk implicitly in Reznek's remarks, but the model-based account presented here has explicated this idea and laid it out in detail. Reznek provides no detail on how he would propose the inference to best explanation to be structured – for example, according to which general inferential strategy is realised in the diagnostic process. Moreover, by discussing the cognitive vehicles supposed to underlie the pursuit of the indicator strategy that is realised in diagnostic inference to the best explanation, by proposing the existence and use of propositional diagnostic models, I have added a layer of detail that is also missing in Reznek, who make no great effort to spell out in detail the process or means of comparison. Hence, it seems that the model-based provides a more detailed answer as to the rationale for psychiatric diagnostics proceeding the way it does.

When we turn to what Reznek's proposal has to offer in terms of the adequacy condition of helping us to understand how the conclusions of the method are supposed to be deemed justified, he again has offered us something. He provides an account to help us grasp the external justification by gesturing towards the relevant science informing our clinical psychopathology, which is what I did. However, when it comes to internal justification, all he has to offer us is that diagnostic conclusions are justified since they are arrived at by an inference to the best explanation, which is also part of my answer. However, since this is all he offers, he seem to miss a relevant aspect of psychiatric diagnostics: exclusion diagnostics, or diagnostic conclusions drawn not because we have an explanation that best explains what the diagnostic condition is, but because we actually have no explanation (no model, as I would say), for the patient's presentation. Such conclusions are justified by identifying which explanations do not apply and then providing a diagnostic label that basically means that the patient has a complaint whose evaluation did not match up with any of our potential explanations of this complaint. This inference and the justification it provides for a diagnostic categorisation of the condition in question is not an inference to the best explanation; it is an inference *qua* the lack of explanation. Although Reznek intends to discuss how diagnostic judgements are internally justified, he apparently missed this aspect of diagnostic practice, or at least his proposal does not address it. By contrast, the model-based account contains an explanation of the inferential work and how it justifies the diagnostic conclusion that has been reached in terms of the inferential pattern of apophatic inferences. Here it seems that the model-based account is preferable over Reznek's as it provides an understanding of how justified diagnostic conclusions – in a class of diagnostic judgements that are not discussed by him although they seem to be present in clinical diagnostics – can be arrived at. The only inferential pattern he puts forward to explain how psychiatric diagnostic reasoning is supposed to arrive at justified conclusions – that is, abduction – cannot account for this class of judgements.

A remaining step to assess Reznek's proposal against mine would be to discuss to what extent it can satisfy the desiderata I set up and showed to be fulfilled by my own proposal. However, as I stated at the beginning of this section, Reznek's intention in his work seems not to have been to provide a full-blown answer to the Methodological Question; rather, he focused on just one aspect of it, the topic of justification. Thanks to this fact alone, his proposal will not fulfil the desiderata. Just think of the desiderata of comprehensiveness and being cognitively realistic: If there is no description of the method of diagnostic reasoning, it cannot be comprehensive and realistic. The same goes for the requirement of helping us to make sense of diagnostic disagreements or the difference between misdiagnosis and malpractice, or of any of the other desiderata I put forward as being preferable in an answer to the Methodological Question. The ideas that Reznek provides are not sufficient to address these issues in a satisfying degree of detail, because his very general point that inference to the best explanation is the inferential basis of diagnostics is not suited to telling us, on the level on which a proposed method would operate, what happens in the case of the phenomena we are interested in. One might suggest, for example, that diagnostic disagreement arises when diagnostic experts disagree about which inference to the best explanation to make once they have gathered diagnostic information about a patient. However, to accept this level of abstractness to account for the desiderata would be a low bar to clear to account for the phenomena pointed out in the desiderata. Saying that we understand what is going on these cases with this level of abstraction would be like saying, if we ask how a biomedical researcher discovers genes responsible for a disorder, that we are satisfied by the explanation, "by induction". Intuitively, few people interested in the topic would be satisfied, and we should not be satisfied with the degree of detail that Reznek's account would provide us with to address the desiderata relevant to achieving a good understanding of psychiatric diagnostics.

In sum, it appears that Reznek's account does not provide a full answer to the Methodological Question. Furthermore, it seems that even in terms of the adequacy conditions for the answer he does provide, his proposal performs worse than the model-based account, given the lack of detail and depth in terms of helping us to understand the rationale behind the diagnostic procedures. This weakness in his proposal for understanding the internal justification of diagnostic reasoning is due to his exclusive focus on inference to the best explanation. Moreover, largely because he does not provide a description of a concrete method at work in diagnostic reasoning, he also fails to fulfil the relevant additional desiderata. Consequently, it seems that the model-based account is more satisfying, as it goes beyond the scope of Reznek's proposal. Even if we set aside the fact that no description of a method of diagnostic reasoning itself is provided by Reznek, the model-based account has substantial benefits over Reznek's account where they address common aspects required for such a proposal.

5.4 Gupta, Potter, Goyer: Interpersonal Knowing

In their paper "Diagnostic reasoning in psychiatry", Gupta, Potter, and Goyer (2019) intend to make a specific contribution to the theory of psychiatric diagnostic reasoning. Their contribution is not a proposal for how to address the Methodological Question, nor any of its aspects. Their contribution, roughly speaking, is a critique of the way that many proposals, which they call cognitive accounts, miss a crucial aspect of psychiatric diagnostics – namely, the role of second-person knowing (i.e., knowledge acquired from the second-person perspective) about the patient for the act of diagnosing in clinical practice. As I will spell out in a moment, they argue that this second-person knowledge is necessary for psychiatric diagnostics. It is necessary since without including such knowledge, a psychiatrist cannot recognise the presence of a mental symptom in a patient. Thus they argue that cognitive approaches to psychiatric diagnostics focusing on the processing of objective data of patients (self-reports and observations) leave out the role of second-person knowing in the identification of psychiatric symptoms. Gupta, Potter, and Goyer would presumably classify my account as a cognitive one, since I do not stress the role of second-person knowledge for the use of disorder-diagnostic models, but rather imply that the relevant propositions are to be evaluated by self-report, observation, and formal testing. If I am right that they would think of the model-based account as a cognitive approach, then considering their argument is worthwhile, because if they were right, the model-based account would be missing something important and would be wrong.

I will argue that Gupta, Potter, and Goyer are not wrong that second-person knowing is crucial in psychiatric diagnostics, but that it is crucial in a different way than they believe – a way that is in fact covered by the model-based account. I will argue that second-person knowing is not necessary for any case of clinical diagnostics to assess the plausibility of the presence of a certain mental symptom in itself, but rather that the right place for second-person knowing in psychiatric diagnostics is a specific aspect of differential diagnostics. Specifically, I suggest that second-person knowing is involved in setting up and testing the diagnostic hypothesis to show that a psychiatric complaint is not a symptom, but rather an unpleasant but normal mental occurrence. In other words, I claim that we need a second-person perspective to argue that perhaps a complaint could be better understood as a non-pathological phenomenon rather than a symptom. However, this perspective is not essential to assess the plausibility of initially considering it as a symptom before comparing it to the alternative non-pathological explanation.

Gupta and colleagues on the other hand claim that the second-person perspective is already necessary to do exactly this, to assess the initial plausibility of a complaint being a symptom in the first place. But before I come to my argument, let me present the ideas of Gupta and colleagues.

Gupta, Potter, and Goyer (2019) claim that the usual understanding of diagnostic reasoning is focused solely on the cognitive evaluation of objective data about patients, which is not sufficient for the context of psychiatry, since a form of interpersonal (second-person) understanding of patients is needed to draw certain diagnostic conclusions in psychiatry. As they put it in their article, they take issue with the idea that diagnostic reasoning

> is a cognitive process involving the manipulation of objective data that takes place in the mind of the individual clinicians. Instead, we argue that psychiatric diagnostic reasoning requires the clinician to use intersubjective ways of knowing even though they are not explicitly acknowledged as sources of evidence in preeminent accounts of diagnostic reasoning. (ibid., p. 51)

They claim that to really grasp the epistemology of psychiatric diagnostic, "a grasp of the role that this kind of knowing plays is necessary." In this sense, "the process of belief formation through second-person knowing is not only what we do but is necessary to diagnostic reasoning in psychiatry because it is a central means by which psychiatrists gather evidence for diagnosis" (ibid., p. 53).[2]

Their approach to interpersonal understanding differs from the previously discussed proposal of Cooper's (2014), because in contrast to Cooper, Gupta, Potter, and Goyer claim interpersonal understanding to be relevant for diagnosing specific symptoms and so to have a proper diagnostic function. If they are right about this, my account would have missed something. Let us look at their proposal.

Their general perspective on psychiatric reasoning is that with a few exceptions – certain neuropsychiatric disorders such as Huntington's disease – psychiatrists make diagnoses by matching elements from the patient's history of illness to sets of operationalised criteria (e.g., *Diagnostic and Statistical Manual of Mental Disorders* [DSM] criteria) so that "[a]part from a clinician simply being mistaken about the correct criteria for a given diagnostic category, it is difficult to claim that psychiatric diagnoses are right or wrong" (Gupta, Potter, and Goyer, 2019, p. 50). From this very abstract commonsensical description of what psychiatrists do in diagnostics, they proceed to the following claim:

> Unlike in general medicine, diagnostic reasoning in psychiatry is less like finding a solution to a puzzle. Instead, it is more like sketching a roadmap that will enable clinicians to understand their patients' problems to identify means to alleviate their distress. The quality of psychiatric diagnostic reasoning must be evaluated in relation to the extent that it facilitates these tasks. (ibid.)

Here, Gupta and colleagues begin to mix up the intrinsic purpose of diagnostics (namely, to identify the present symptoms and disorders) with practical purposes that diagnosis serves in psychiatry, namely treatment selection. The result in the passage just quoted is that they make a statement about the *purpose* of diagnostics ("being a roadmap") rather than about its *nature* ("finding a solution to a puzzle"). Why we should believe that psychiatric diagnostics, as opposed to diagnostics in other medical fields, should be thought of along these lines remains unclear. Instead, they go on to point out that a grasp of the patient's problems that would feed

2 However, Gupta, Potter, and Goyer are inconsistent (or at least unclear) about how important second-person knowing really is in psychiatric diagnostics. While in the passages quoted here it sounds like its presence is ubiquitous and generally necessary, later in their paper they make more modest claims, such as "intersubjective knowing is not merely a helpful add-on to subjective or objective knowing, but *in some cases* forms an integral part of knowing a person" (2019, p. 57; my emphasis).

into their "roadmap" requirement for diagnostics requires an interpersonal understanding, an insight in terms of the psychiatrist's own imagination and experience of what the person is going through:

> Constructing an accurate roadmap of a patient's psychiatric problem seems to require more than the kinds of objective data about the person that serve as evidence in support of most medical diagnoses. Understanding – or even accurately describing – a person's mental state including her thoughts, feelings, and experiences is intersubjective; that is, it requires an awareness of the patient's world that is mediated by the clinician's own thoughts, feelings, and experiences when in relationship with the patient (Pauen, 2012). (Gupta, Potter, and Goyer, 2019, p. 50)

Elaborating further on the idea of the relevant kind of interpersonal knowledge, they argue as follows:

> [B]ecause the clinician does not have direct access to the patient's mental states (such as his or her beliefs, emotions, desires, motivations, and meaning making), the clinician needs to draw on resources such as imagination and empathy, and to continually confirm one's inferences with the patient while adjusting her understanding of how the patient's world is experienced by the patient himself and noting how the patient shifts and adjusts to the clinician as well (cf. Pauen, 2012). (ibid., p. 54).[3]

After presenting their view on psychiatric diagnostics and the importance of interpersonal or second-person knowing, they illustrate their case with examples that all attempt to drive home the same point in a similar fashion. Let us look at one of these examples: the diagnosis of major depression. Regarding the diagnosis of major depression, they claim that "The criteria set contains some items that can be identified strictly subjectively (e.g., diminished interest, fatigue, feelings of worthlessness) and some that can be assessed objectively (e.g., 5% weight loss). There are no

3 Gupta, Potter, and Goyer (2019) repeat the point once more in terms of Gallagher's (2009, p. 290) notion of "participatory sense making". They paraphrase Gallagher as arguing, first, that "for me to understand how you experience your world, I need to attend, imagine, empathize, and listen with openness to your ways of indicating what it is like to be you and how you make sense of your world" and, second, "that I need to respond to your communications and behaviors with an eye toward clarifying, correcting, offering possible explanations, inquiring more, and seeking opportunities for emotional connection" (Gupta, Potter, and Goyer, 2019, p. 55). They conclude that "making sense of our interactions and relations with others, therefore, seems to require second-person knowing" (ibid., p. 55), which is the kind of knowing discussed in the previous quotes.

items that are explicitly intended to be known intersubjectively" (2019, p. 57). However, they go on to claim that "if we examine certain items more carefully, intersubjective knowing must be at play in their assessment" (ibid., p. 57). To demonstrate this, they pick out the symptom of depressed mood: "Depressed mood most of the day, nearly every day, as indicated by either subjective report (e.g., feeling sad or empty) or observation made by others (e.g. appears tearful)" (APA, 2013, p. 160).

While Gupta, Potter, and Goyer acknowledge that tearfulness as an observed behaviour might be an objective indicator, the problem remains that "tearfulness may also indicate other mood states, such as anger, anxiety, frustration, or joy" (2019, p. 57). They argue that to adjudicate between these possibilities, second-person knowing is needed:

> it seems as though intersubjective knowing is required to interpret the objective observation of tearfulness. Some examples that would support the hypothesis of sadness in the presence of a tearful patient (who is not subjectively reporting depressed mood) might include the telling of life experiences that the clinician finds sad ("finding sad" requires empathy or imagination), and the clinician's own feelings of sadness in the presence of the patient and that patient's life events (which requires emotion). Additionally, behavioral gestures such as a downcast gaze may also provide evidence that the patient is depressed, but this again requires an interpretation of behavior that could be consistent with other emotional states. In other words, although depressed mood can supposedly be assessed in objective terms (seems to be tearful), intersubjective knowing is needed to act as an intermediary between the third-person observation and the first-person state of depressed mood. (ibid., p. 57)

What Gupta and colleagues are thus arguing is that it is valid to make a judgement about the presence of "depressed mood" based on introspective report of things introspectively associated with depressed mood, such as feelings of emptiness or sadness or observation of behavioural features such as tearfulness. However, tearfulness alone as an objective behaviour is not enough, they claim, since it may be caused by mental conditions other than depressed mood. Therefore, knowledge that in their view is second-person knowledge must be generated in order to make plausible the interpretation of tearfulness as indicating depressed mood – to assess whether, in light of biographical details and the interaction with the patient, it is plausible that the patient is indeed sad on the empathetic level. This case is supposed to show that second-person knowledge is necessary to make a supposedly possible diagnostic judgement (depressed mood, based on behaviour or tearfulness) plausible.

Another example discussed by Gupta and colleagues concerns generalised anxiety. One of the criteria of generalised anxiety is "[e]xcessive anxiety and worry (apprehensive expectation), occurring more days than not for at least 6 months, about a number of events or activities (such as work or school or performance)" (APA, 2013, p. 222). Regarding this symptom, they claim:

> Although the state of being worried can be reported subjectively by the patient, whether or not the worry is excessive is more complex. For the physician to judge excessiveness she may first seek out some objective data (e.g., time spent worrying), but similar to the behavior of tearfulness in the depression example, the physician needs a method to make the jump from a certain quantity of worrying to a judgment of excessive. Such a judgment requires understanding of this patient given his personality as well as the context, content, and preoccupation of his worries. [...] Such an assessment cannot be objective in the sense that there is no true amount of worrying that is the correct amount for a given person's situation. In other words, there can be no recourse to an objective assessment that will not eventually loop back to an intersubjective assessment. (Gupta, Potter, and Goyer, 2019, p. 58)

Regarding another potential feature of generalised anxiety, namely irritability, they claim that that although it might seem at first glance that this feature might only be known by self-report, this is not the case:

> a person may not endorse irritability, yet the clinician finds that the patient is behaving in an irritated manner in the clinical encounter. It may be that the person does not generally feel irritable, but is feeling irritable toward this psychiatrist at this point in time. However, it may also mean that the person does not understand what irritability is, or does not wish to acknowledge his irritability. To make this determination, the clinician would need to engage in a full range of strategies of knowing the patient to evaluate the credibility and plausibility of the self-report including asking for a more detail behind the subjective report (how the patient is feeling at the moment), using his imagination (how the patient is perceived by others), and trying to establish a shared language to describe the patient's feelings based on what is being discussed and interpreted between them. (ibid., p. 58)

In other words, whether the patient is irritable will again depend a complex set of information, assessing which supposedly requires second-person knowledge of the patient since otherwise one may neither judge the behaviour of the patient to indicate irritability nor be sure that self-reports of present or recent experiences and behaviours indeed indicate irritability. I offer a last quote here that, though made in the context of the depression example, seems to speak to all these examples interchangeably: "This illustrates Pauen's point, as noted, that objective knowledge (of a patient's sadness, based on the observation of tearfulness) needs to be grounded in some prior second-person knowing" (ibid., p. 58). Now that I have presented the ins and out of Gupta et al.'s proposal, let me turn to assessing it.

Guptas and colleagues' proposal is modest. They do not attempt to provide a full approach to diagnostic reasoning, but rather claim that mainstream approaches to diagnostic reasoning focus on the cognitive processes taking place, and that there is a necessary aspect of diagnostic reasoning that all these cognitive approaches miss. This aspect is that identifying the presence of a psychiatric symptom requires the use of knowledge that can be gained only from the second-person perspective. I assume that they would classify my approach as a cognitive approach that misses this component. Given this assumption, they would claim that something essential is missing in the model-based account. Hence my account would be wrong.

My response will be to argue that Gupta, Potter, and Goyer's (2019) argument fails to establish the necessity of second-person knowledge in the identification of mental symptoms, and that there is therefore no reason to assume that the model-based approach, or any other cognitive approach, fails because of its absence. My basic argument for this is that Gupta and colleagues overstate the epistemic role of second-person knowing by exaggerating the irreducibility of the second-person perspective in a way that does not align with actual claims made in the original sources they use, namely Michael Pauen's work. This is a problem, since instead of offering their own independent arguments for their claims, they rely repeatedly on Pauen as an authority to justify their claims, and suggest that their positions are paraphrases of his. Without Pauen, there is no reason to believe them. When we clarify what Pauen's actual position is and apply it to the context of psychiatry, the irreducibility claim made by Gupta and colleagues collapses, and with it their argument for the necessity of the second-person perspective in diagnostic reasoning. Hence, they fail to show the inadequacy of cognitive accounts including the model-based account. After presenting this principal argument for why they have not established the necessity of second-person knowing in psychiatric diagnostics, I will take one of the examples they provided to illustrate their argument and show why what was shown in principle can also be shown in practice – that is, I will also show why second-person knowing is not necessary in the specific case.

To supplement my criticism of their argument for why second-person knowing is necessary to identify a psychiatric symptom, I will point out the important role that second-person knowing has – a role that does make it highly relevant in psychiatric diagnostics. More specifically, second-person knowing is required in the context of differential diagnostics. Here, however, it does not contribute to identifying whether the patient meets what is required to have a certain mental symptom. Rather, it contributes to our folk-psychologically informed considerations as to how it might be that the patient has a certain distressing mental state or disposition for reasons that are not psychopathological or in other ways medical. In other words, the second-person perspective and second-person knowing do not come in when learning something about the patient that allows us to attribute a symptom to them; rather, they contribute to the psychiatrist's capacity to recognise when a patient's complaint *is not* a psychiatric condition. This role, however, as I will discuss, is covered by the model-based account. Thus, I show that Gupta et al.'s specific argument for second-person knowledge fails, and that the perhaps intuitively plausible idea that the second-person perspective must play a role in diagnostics is not wrong, but, if considered correctly, is also no threat to the model-based account. Let me begin by clarifying Michael Pauen's understanding of the second-person perspective and second-person knowledge.

The reference to Pauen's (2012) paper "The second-person perspective" at multiple points in Gupta, Potter, and Goyer's (2019) paper to support the irreducible relevance of interpersonal knowing in diagnostics is curious. In his article, Pauen discusses different epistemic perspectives for gaining epistemic access to mental states, which he roughly divides into first-person accounts (introspection), second-person accounts (interpersonal knowing), and third-person accounts (objective data). He argues, amongst other things, that our third-person access to mental states depends on our second-person access in an irreducible way. Crucially, Pauen sums up the irreducible relevance of the second-person perspective in relation to the third-person perspective in two regards. First, he highlights its importance for the initial calibration of our third-person access to mental phenomena:

> the second-person perspective is needed in order to *ground* third-person claims regarding mental states. This is why it cannot be reduced to the third-person perspective. If we want to identify the neural correlates of, say, pain in an experimental subject, we have to make sure that the experimental subject really *is* in a pain state in the first place. Doing this requires the application of the relevant concept, that is, the concept of pain. As we have seen above, employing mentalistic, particularly phenomenal concepts like "pain" implies that the speaker is able to simulate and ascribe the mental state in question. And this just *is* ascribing a mental state from the second-person perspective. (2012, p. 45)

In other words, we as people conducting science or at least attempting to objectify ways of attributing mental states to other need to decide in the first place to whom we are willing to attribute a certain mental state. To calibrate, for example, an MRI method or a questionnaire to recognise a certain metal state or disposition in someone, I first need to determine whom I will take to be in this state. The second-person perspective is therefore a means to calibrate, to ground, my third-person method.

The second form of relevance of the second-person perspective is in making third-person ascriptions of mental states interpretable or understandable: "the second-person perspective is also needed in order to understand third-person claims regarding mental states. Again, the reason is that using mentalistic, particularly phenomenal concepts requires an imagination or simulation of the mental state in question" (Pauen, 2012, p. 46).

What Pauen seem to be saying here is that the other way the second-person perspective is important is in the understanding of what it means to make a claim that a person is in a certain mental state. The reason is that if we attribute a mental state, especially one that entails an experiential dimension, to someone, fully grasping what it means to attribute this mental state requires that we can imagine what it is like to be in this state in terms of experience and dispositions.

How do Gupta and colleagues employ Pauen to defend their ideas regarding the irreducibility of the second-person perspective? It seems that they believe that in *any* instance of diagnostic reasoning, both the *grounding* and the *understanding* aspects need to occur in order to support the clinician's diagnostic reasoning. This means that every time we ascribe a mental symptom to a patient, we have to do two things. First, we must attribute this mental state to them exclusively by means of our second-person ways of reading someone else's mind in interaction with them, as we do every day in a folk-psychological manner (i.e. grounding). And second, we must have simulatory access to the mental states we attribute, to fully grasp what these mental states are that we are attribute to the patient (i.e. understanding). Yet if we read Pauen carefully, it is not necessary that both aspects must be present in every instance of attributions of mental states.

As discussed above, Pauen believes that the understanding portion that comes with the second-person perspective is generally irreducible if we want to fully grasp what we are talking about in attributing mental states to others, but he does not believe that the grounding portion is irreducible in any instance of attribution. Rather, he believes that some second-person attribution of a mental state is necessary to start with, but that later on, an alternative tracking method calibrated on such attributions may well substitute for the second-person grounding of an attribution. In other words, attribution can perfectly well take place from a third-person perspective once a way to do so has been established. As Pauen himself states very clearly:

> third-person perspective taking is definitely possible, even with respect to mental states like feelings, beliefs, and desires. This is, by the way, what we have to expect given that perspectival differences are differences on the level of epistemic access, not on the level of epistemic objects. If this is so, then it should be possible to take different perspectives on one and the same object – as it is the case with respect to the third- and the second-person perspective regarding mental states. (2012, p. 46)

As Pauen emphasises, the difference between the perspectives is not their *epistemic object* but a difference in *forms of epistemic access*. The very same object (the mental state X) may be epistemically accessed by introspection, intersubjective knowing, or a third-person method calibrated on our otherwise attributed mental states. Epistemic access *qua* third-person perspective does not presuppose the employment of a second-person perspective to ground it every time anew but only in its calibration phase. What conclusions can we draw from this first comparison between the claims of Gupta and colleagues' and Pauen's positions?

Straightforwardly, it appears that Gupta, Potter, and Goyer may rightfully base on Pauen's account their claim that the *understanding* portion of the second-person perspective (having an empathetic understanding of what it is the patient is experiencing) is indeed irreducible. But they cannot justifiably employ his approach to support the claim that any attempt to diagnose a mental symptom must be grounded in second-person understanding, because it would be perfectly coherent to use only a third-person method that was calibrated on second-person attributions. As a result, all that Gupta and colleagues can claim is that for a clinician to understand what it is like for a patient to have a certain mental symptom requires the understanding portion of the second-person perspective, and requires that at some point the method by which psychiatric symptoms are assessed has been grounded in the second-person perspective. If this is the case, however, attributions can correctly be made *in situ* when previously grounded with a second-person approach, without taking the second-person perspective into account every time they are made. Hence the irreducibility claim that was meant to extend to each diagnostic attempt crumbles to the necessity of some grounding in the past. It amounts only to the necessity of making an interpretation of this attribution *qua* emphasising, in order to know what it means for a patient to suffer from a certain symptom. However, even these two remaining necessities of the second-person perspective face problems if we attempt to apply Pauen's ideas not to normal psychological phenomena almost everyone knows from first-hand experience – like beliefs, desires, or pain – but to the context of psychiatric phenomena.

If we do not just consider how the claims of Gupta and colleagues hold up in light of Pauen´s thoughts on the second-person perspective, but also consider how applicable Pauen's approach is to psychopathological phenomena rather than normal psychological phenomena, it seems that even the remaining necessity fortresses that Gupta et al. could defend turn out to crumble. First of all, the idea that all third-person access must have been grounded at some point on second-person access has seemed plausible so far. However, if we look at psychopathological phenomena, this starts to seem problematic. In contrast to the mental states that Pauen discusses in his paper, which are common propositional attitudes and phenomenal experiences, at least some psychopathological phenomena seem hard or impossible to empathise with in the way required when taking Pauen's second-person perspective. This requires that we arrive at the attribution of the mental state by drawing on our own experiences with this very state and assuming the other to be in the same type of state. As we discussed earlier in the context of Cooper's proposal, it seems hard or impossible to put ourselves is a valid imaginary perspective that gives us a good grasp on what it is really like to have intrusive memories, be in pre-psychotic prodromal state, or have a Capgras delusion or some other severe mental condition if we have never experienced these things ourselves.[4] For at least some psychopathological conditions, grounding plausibly must have taken place based not on the second-person perspective but on first-person information from patients acquired in third-person forms such as via verbal reports, behavioural observations, and potentially formal cognitive or biological testing. If we do not need second-person access in grounding, then it seems that the second-person perspective is not necessary at all for developing and engaging in the diagnostic reasoning required to diagnose various psychiatric mental symptoms. Therefore, the necessity claims regarding this first aspect of *grounding* and the second-person perspective in diagnostic reasoning seems dispelled. What about the other aspect, understanding?

If what I argued in the last paragraph is correct, we can derive from this another interesting point that speaks against Gupta et al.'s claims. We have already established that a second-person understanding is not a necessary part of any *in situ* diagnostic procedure, since it is possible that such procedures (however they are grounded) may be carried out based entirely on calibrated third-person methods, such as taking into account the self-reports of patients, observing their behaviour, or doing some sort of testing. However, if there are cases in which the *grounding* could not be done in a second-person form for at least some mental symptoms, it seems that there could also be no second-person understanding of these conditions in the context of diagnostics. This means that, in contrast to what we would expect according to Pauen regarding the use of mental terms, when we say that patients suffer from these symptoms without second-person grounding, we are not saying so with a clear empathetic take on what it would be for us to be in this state and therefore understanding this diagnostic label through second-person access. It would then follow that second-person understanding in psychiatric diagnostics is not only not necessary but in some cases is even impossible.

In the last few paragraphs, I have argued plausibly that Gupta, Potter, and Goyer (2019) fail to show that second-person knowledge is essential to psychiatric diagnostics based on their adoption of Michael Pauen's account of the second-person perspective. More than that, I have shown that if we take Pauen's account and attempt to apply it to psychiatric phenomena rather than typical mental states, it even

4 A problem that, as one may add, is today commonly accepted in psychiatry and one of the drivers to include individuals with the lived experience of psychiatric disorders on almost any levels of mental health care, instead of relying solely on people lacking these experiences and imagining how thing are for these patients (see, e.g., Fusar-Poli et al., 2022; Happell et al. 2022; Sunkel and Sartor, 2022).

appears that in some cases it is plausibly not even *possible* to employ the second-person perspective. The irreducible role of the second-person perspective in psychiatric diagnostics that Gupta and colleagues attempted to make plausible therefore seems a claim that remains unproven. Although their point is rebutted in theory, however, one may still think that their examples make a persuasive point that cannot be put aside by a principled argument. Do they not have a point with their examples that plausibly generalises? To address this worry, let me next take one of their cases, the diagnosis of depressed mood, to show how their examples can be deflated as well.

In their example of depressed mood, Gupta and colleagues identify two principal approaches suggested in the DSM to ascribe a depressed mood to a patient. The DSM states that depressed mood is "indicated by either subjective report (e.g., feeling sad or empty) or observation made by others (e.g., appears tearful)" (APA, 2013, p. 160), and they assume that according to the DSM, depressive mood might be diagnosed based on either self-reports or observations. However, in discussion of diagnosing depressed mood by observation, through tearfulness, they seek to find an implicit route via which to back up the necessity of the second-person perspective: "tearfulness may also indicate other mood states, such as anger, anxiety, frustration, or joy" (Gupta, Potter, and Goyer, 2019, p. 57). Thus, to really determine whether tearfulness indicates depressed mood requires more – for example, "telling of life experiences that the clinician finds sad" and "behavioral gestures such as a downcast gaze" (ibid.). Although this information can be assessed objectively, it is lent support only thanks to the second-person perspective, since "intersubjective knowing is needed to act as an intermediary between the third-person observation and the first-person state of depressed mood" (ibid.). Though initially plausible, there are severe problems with this approach.

The first problem is that Gupta and colleagues employ an artificial interpretation of tearfulness. It is true that someone may cry if they are angry, sad, or happy, but just having tears in your eyes crying is not all that the usual thought be entailed by tearfulness. Just as the word *jubilatory* does not only entail that is uttering a laconic "YEY!" but also evokes expectations about other behaviours, tearfulness evokes a certain overall expectation. This expectation would include certain body language (e.g., drooping shoulders, shakiness, downward gaze, motor retardation) and speaking behaviours (speaking more quietly, slowly and hesitantly or with a shaking, raspy voice in an almost logorrheic manner). Of course, there is ambiguity, and it is fine to say that someone is tearful if, for example, she has just won Wimbledon, raises her arms, and screams 'Yes!' with tears in her eyes. However, it seems that this additional qualification (screaming 'Yes!', raising arms, and having won Wimbledon) is necessary to prevent the initially described associations we have with tearfulness from coming to mind when the word is used. If this is true and tearfulness, despite its ambiguity, commonly has a primary meaning (the one I proposed above) in the sense that it is the first thing we commonly think of when we think of "tearfulness," it appears fair to also assume that this unqualified meaning is also intended in the DSM, which attempts to be pragmatic and concise in most of its descriptions.

In other words, on a common interpretation of tearfulness, everything that Gupta and colleagues claim to be implicit aspects of what a psychiatrist looks for thanks to a second-person grounded approach is already entailed in the use tearfulness in the DSM. No personal interpretive grounding work based on empathy would then be necessary, just a proper look at the patient. However, not everything that is important for interpreting tearfulness as an objective indicator of depressed mood is covered in this way. What about the relevance of empathy with the patient's sad life experiences, which arguably do not fall under "tearfulness" but are an object of second-person interpretation? Even if all these behavioural aspects are captured by "tearfulness", this dimension is not, and it may well be crucial to interpreting the patient's tearfulness. This brings us to the next problem.

The second problem with Gupta, Potter, and Goyer's proposal is that they as quoted earlier proclaim that to identify patients' tearfulness as an indicator as an objective sign of depressed mood the clinician would need to have to use information about occurrences in the patient's life that the clinician themselves finds sad. However, if the patient's emotional reaction appears to be fully intelligible because it apparently is the result of an event that would have made sad almost anyone (i.e. you loved mother died three weeks ago) this rather seem to speak for the tearfulness to be an expression of a normal state of sadness that is easy to emphasize with and not a sign for a pathological state of depressed mood, so that being well explainable in the context of a sad life event would rather (or at least as well) be point for the differential diagnose of normal sadness rather than depressed mood. That these two things are different, and therefore should also be kept distinct, seems apparent if we judge by existing phenomenological work on depressive mood (e.g., Ghaemi, 2007; Ratcliffe, 2015). To show this let us look at an exemplary description of the depressed mood, in an extract of a description provided by a patient:

> All connections are lost. One feels or is like a little stone, lost in the endless grey of a fading landscape. The sensation of smallness, insecurity and loss can become so strong, that one almost has a feeling of a dream world in which even being oneself is anything more than an abandoned point, like a dried leaf moved here and there in a lifeless autumnal world. [...] The solitude of the depressed is different from every other solitude and from every other state of abandonment. One is not alone in a house, in a city or country. For the house is like lost, it does not mean protection anymore; the city is not a familiar city, the country is not homeland anymore, the starry sky burnt by the ice... However, now one is not humans in the flesh, with heart, strengths and spirit to bear solitude–one is a stone. A stone that suffers and thinks; something like that exists. So to speak, one is retro-evolved in stone. Sometime I have thought, "Now I know what is like to be a stone". It is even too clear that this little stone in the cold universe, this enigmatically afraid and doubting man

> strives to grab himself, with ineffable, fervid effort, and find a hold in everything on which he can in some way grab himself (human, animals, things)... […] What is left of the human, when he is deprived of the rational capacity, the intuitive force, the capacity of transmitting and receiving love? A little intellect is left...it is nothing but the bed of a dried stream, a binary on which nothing travels anymore. It is in himself a poor dried leaf. […] It does not matter which fuel you put into the furnace of suffering and for which reason the fire develops. In a sense it is a good that objects are found, even though this sharpens the suffering; because the true and horrible essence of anguish, in the depression, is its lack of an object. (Tellenbach, 1980, pp. 250–252)

It appears that such an experience goes well beyond and is very different from normal sadness. In line with Cooper's (2014) considerations discussed earlier, we would not expect someone unfamiliar with such an experience to be able to properly simulate it in their mind as the expected mental consequence of sad life events. Thus, against Gupta and colleagues, it seems that using the second-person perspective in the context of diagnosing depressed mood is an epistemically problematic move. A clinician following the ideas of Gupta and colleagues, who has never experienced depressed mood and models what he attributes based on their experience of sadness, would, if they meet a depressed patient, wrongly attribute to him sadness and just call it "depressed mood". Also, if they meet someone who experiences something sad and whose tearfulness is fully intelligible in the context of their experiences, they will end up telling them that they are depressed. This will not always be wrong, as many people who experience depression have had sad experiences; however, often they will be wrong because many of us have sad experiences and are tearful, but seemingly few of us at the same time make the experience of depression described above. Many people are simply sad. Hence, the clinician would often end up wrongly telling people they have a depressed mood when there is actually only sadness. As a result, assessing the presence of depressed mood based on evaluating behaviour focusing on the patient's life story, as proposed by Gupta and colleagues, seems to be mistaken and should be discarded.[5] Given the two problems with the

5 One objection I might predict is that it seems that if one indeed assumes that the term "tearfulness" is meant to cover all the observable aspects taken to be associated with depressive mood, and the understanding approach to depressive mood based on life events fails, would that then mean that in psychiatric diagnostics the appearance of tearfulness with all its aspects is indeed treated as sufficient to diagnose a mood as rich and multifaceted as that described by Tellenbach's patient? That seems to be quite an epistemic leap.
My response to this worry is twofold. First, although the manual intends the diagnosis of depressed mood to be possible by self-report only or by observable behaviour only, its seems that what we would want for a diagnosis is self-report and behaviour both suggesting this symptom. And indeed, in a clinical context we will almost always have both kinds of information: most patients tend to speak about their experiences and suffering when they enter

example of depressed mood provided by Gupta and colleagues that I pointed out in the last few paragraphs, it appears that the second-person perspective is not needed to assess depressed mood. Even worse, it might lead the clinician to wrongly assume that they would be able to adequately empathise with what the patient is experiencing. Having provided principled reasons why Gupta and colleagues fail to establish their claim regarding the irreducibility of second-person knowing in diagnostics, and having also provided an exemplary demonstration how one of their examples in support of their case fails, let me come to something more constructive. I will now point out where second-person knowing is indeed crucial in psychiatric diagnostics, though in a very different way than that argued for by Gupta and colleagues, and in a way that is covered by my own account.

While Gupta, Potter, and Goyer's (2019) argument aimed to show the necessity of second-person knowing in diagnostics, more precisely its necessity in assessing specific psychopathological symptoms, there is another area of psychiatric diagnostics covered by my approach in which it plays a role. Second-person knowing comes into the picture when the psychiatrist starts to consider potential diagnostic evaluations of the presented complaint in which, rather than being a psychiatric symptom or a non-psychiatric medical symptom, it could also be a psychological complaint without any symptom value – that is, a psychological phenomenon falling into the scope of normal psychology rather than psychopathology or other areas of medicine.

---

a diagnostic setting. If a clinician finds himself forced to make the diagnosis based on only one information source, be it on self-report (e.g., because he works for a telehealth service) or only by behaviour (e.g., because the patient suffers from mutism and so cannot speak to the psychiatrist), he may have to consider the tearful appearance of the patient alone. However, to justify using tearful behaviour as valid evidence to diagnose depressed mood, he will make sure to have a better basis than just the momentary assessment of the patient;s appearance. First, the psychiatrist will ensure that the preamble of the diagnostic criterion is met: that there are reliable reports that the patient has been in this behavioural state most of the time for at least two weeks. And second, the psychiatrist will consider possible differential-diagnostic options. For example, he will assess whether the patient may have started to take medication that is associated with side-effects such as the development of psychological complaints of depressed mood (e.g., some hormonal contraceptives; Skovlund et al., 2016; Mu and Kulkarin, 2022), such that the timing of the onset of the complaint may better be understood as a psychological side-effect of medication rather than as the symptom of a psychiatric disorder. Only if the criterion is fully met regarding the timespan of tearfulness, and the information gathered about the patient does not better support a differential-diagnostic reason for the patient's presentation, may the psychiatrist provisionally conclude that the patient suffers from depressed mood. However, it should be noted that even with all these aspects in check, my sense is that most psychiatrists would be rather uneasy about making this diagnosis without self-report, and would be eager to get such self-reports from the patient as soon as possible.

If we recall my examples in the second chapter of this thesis, we may think of the example of the complaint alogia. I presented a range of potential ways in which it may be assessed, leading to various outcomes. I showed that the initial complaint may turn out to be a psychiatric symptom under some circumstances but may also turn out to just be a normal psychological phenomenon, such as a hesitation to speak to the psychiatrist out of worry about receiving a diagnosis. In the latter case, the diagnostic procedure relies on a propositional qualitative model containing a set of propositions that, if they all applied to the patient, would together indicate that the reluctant speech of the patient is not a pathological problem. This model, however, was not based on any scientific background knowledge, but on the folk-psychological belief–desire–motivation psychology that we use in everyday contexts. Unless this model is acquired by learning it from another clinician or via the literature (which is the case often enough throughout clinical education), it may be that the clinician comes up with such a model based on considerations using their own capacity to empathise. In this case, they might ask themselves, "Based on my initial idea of who they are, what might be a reason for this person to be so reluctant to talk openly to a clinician?". But again, no case of *in situ* grounding is necessary, even though plausibly it stands at the beginning of all such models and may be entertained to initially develop them. So, the modest role of second-person knowledge, according to my account, is as the original basis, and perhaps sometimes the *in situ* grounding, for the assessment of how plausible it would be to consider a patient's complaint to be a non-pathological psychological phenomenon.

I conclude my discussion of Gupta, Potter, and Goyer (2019) by saying that the second-person perspective does not seem to be irreducible in psychiatric diagnostics in the sense they claim. Hence, they fail to show that cognitive accounts to diagnostic reasoning (a label under which my proposal might fall) miss something crucial in the assessment of symptoms if they do not acknowledge the centrality of the second-person perspective. Gupta and colleagues therefore do not endanger the plausibility of the model-based proposal. Moreover, I have indicated why I believe that the second-person perspective is indeed central to a different aspect of diagnostics, namely the clinician's attempt to understand a patient's experiences and behaviour in a non-pathological sense, which I discussed in more detail in Chapter 2 and Chapter 3.

5.5 Fuchs and Parnas, Sass, and Zahavi: The Phenomenological Proposal

The final alternative philosophical position that I will discuss is one held by researchers in the field of phenomenological psychiatry, a tradition stemming from the ideas of Husserl (1900) and first applied to psychiatry by the philosopher-psychiatrist Jaspers (1913). Among authors inspired by this tradition, some directed their attention towards psychiatric diagnostics. Some well-known figures amongst them are Fuchs (2010) and Parnas, Sass, and Zahavi (2013). For reasons of simplicity, I will call their related positions "the phenomenological proposal".[6]

The basic idea of this proposal is that psychiatrists, when encountering the patient directly, recognise the gestalt of the present disorder in the patient's presentation, unmediated by previous recognitions of symptoms. This proposal is presented in several short remarks. There is no explicit indication as to whether it is intended as a comprehensive philosophical account to psychiatric diagnostics or whether, as in the case of Reznek, it is intended to limit itself to only some aspects of diagnostics. Be that as it may, if one decides to read them as attempting to provide a full proposal, my criticism would be that the proposal is incomplete and that what they supply is inconsistent with my proposal, which is no problem for my account, since I will show that their proposal is implausible. One may also read them as intending only a limited account, most plausibly providing the infantile pattern at work in psychiatric diagnostics, and thus aiming to provide something that meets the adequacy condition rationalising the diagnostic procedure. In this case, their claim would still be incompatible with my proposals regarding this point, and I will argue that their account is implausible. Moreover, I will argue that if their proposal were right, any full-blown account of psychiatric diagnostics would deprive itself of the possibility of fulfilling several of the enumerated desiderata, since their assumed diagnostic pattern makes it impossible to address them. Before I provide my argument, however, let me present their account. To outline their proposal, I will begin with the remarks of Fuchs.

In his article "Subjectivity and intersubjectivity in psychiatric diagnosis", Fuchs (2010) presents a general and affirmative approach to what he thinks experienced psychiatrists do when they diagnose psychiatric disorders. He claims:

> experienced clinicians do not diagnose and practice by ticking off the diagnostic criteria of the manuals. They work with the prototypal approach to diagnosis […] that help[s] to grasp the essence of a phenomenon as an organizing and meaningful "gestalt" over particular details. (ibid, p. 271)

Fuchs does not provide details of why and how the process of the direct recognition of a clinical gestalt is supposed to take place in diagnostics. Parnas, Sass, and

6 My interpretation of the phenomenological proposal, as well as two of its problems (its relationship to pattern recognition and critical reasoning) discussed in this subsection, have previously been developed by me in Kind (2023). I reiterate these points here as part of my extensive discussion on the phenomenological proposal. Without including them, I would not be able to present a comprehensive picture of the phenomenological proposal and its weaknesses for comparison with my model-based account.

Zahavi (2013), however, provide an account of why diagnostics is supposed to directly address the whole gestalt of a disorder in their paper "Phenomenological psychopathology and schizophrenia: Contemporary approaches and misunderstandings", where they discuss the nature as well as the epistemic access to the clinical gestalt in more detail.

In discussing the nature of a disorder's clinical gestalt and the epistemic constraints it puts on the possibilities of how one may recognise a patient's psychopathology, they state that the clinical gestalt of a disorder is "not a simple aggregate; [as] the 'whole is more than the sum of its parts.' This unity [of the gestalt] emerges from the relations between component features and is influenced by the whole (part-whole relations)" (ibid., p. 275). Here, the "components" are symptoms of mental disorders. In other words, the occurrence of whole clinical gestalt is the result of some sort of interaction effect (therefore "more than the sum") of the presence of all the relevant components (i.e., symptoms) at once, which gives rise to the clinical gestalt of the disorder.

With regard to epistemic access to the clinical gestalt of the disorder and its symptoms, they claim that "[a]spects of a Gestalt [...] may be focused on in diagnosis or research; but one must remember that these aspects are interdependent in a mutually constitutive and implicative manner" (ibid.). They go on: "What, then, defines a given individual experience/expression as a specific symptom or sign, [...] articulates itself from within an experiential expressive whole [of the gestalt]" (ibid.). What Parnas, Sass, and Zahavi seem to be saying here is that while it may be possible to focus on single aspects of the clinical gestalt, this is possible only if at the same time the whole clinical gestalt is also recognised. Therefore, while the clinical gestalt and its components are mutually constitutive in their presence, it is the gestalt enjoys epistemic primacy in that, according to Parnas and colleagues, it is only in the context of this gestalt that symptoms "articulate" themselves – that is, can be singled out. Next, let's try to put together what Fuchs and Parnas and colleagues have offered us.

According to Fuchs, the psychopathological feature recognised first and attributed to a patient is the whole gestalt of a disorder, existing over and above any of its details. This means that the psychiatrist does not first discern symptoms and signs, but rather directly recognises a disorder based on the prototypical gestalt as it shows up in the patient's behaviour and reports. Hence, the first step of the diagnostic reasoning process is that the psychiatrist directly recognises the disorder (e.g., a major depression). This point of Fuchs's recurs, though with a little more explanation of why this has to be the case, in Parnas and colleagues' argument. In the first quoted passage quoted above, they tell us that the gestalt emerges with its components (the symptoms) to the clinician. But at the same time, they make clear in the second quoted passage that there is no way to get a valid grasp of these components other than the gestalt. With this claim, they attribute a kind of epistemic primacy to the disorder gestalt in relationship to the symptom, as there seems to be no way around starting by grasping this gestalt if one is attempting to grasp the single symptoms. While Fuchs claims the epistemic primacy of the gestalt merely descriptively (this is just what psychiatrists do), Parnas, Sass, and Zahavi provide an explanation of why this is the case.

Merging Fuchs's and Parnas and colleagues' remarks provide us with an outline of how the phenomenological account of diagnostic reasoning is supposed to work, but details about how the initial recognition of the clinical gestalt is supposed to take place are sparse. However, one interpretation of what the phenomenologists have in mind suggests itself in Fuchs's and Parnas and colleagues' use of the terms *prototype* and *gestalt* – namely, that psychiatrists engage in a form of pattern recognition, prototype processing, that leads to the recognition of the clinical gestalt of a disorder. Let me explain.

The prototype theory of pattern recognition in cognitive psychology is a model of pattern recognition according to which different prototypes of objects are memorised by the system:

> in the process of pattern recognition, outside simulation only needs to be compared with the prototype, and the sense to objects comes from the matching between input information and prototype. Once outside simulating information matches best with a certain prototype in the brain, the information can be ranged in the category of that prototype and recognized. (Pi et al., 2008, p. 435)

An essential feature of this kind of pattern recognition is that it contains top-down processing and no bottom-up processing (ibid., p. 436). Recognising the relevant *object begins with the matched prototype itself*. There is an immediate matching between information input and prototype, rather than an intermediate step in which aspects of what will be identified as a prototype are first recognised independently and then found to constitute a prototype, which would be a bottom-up process. This description seems to match well with the idea of a direct and unmediated recognition of the disorder gestalt, which may later be discerned in its constituent elements. But does the notion of a prototype understood along these lines fit with the idea of the gestalt? We might just take Fuchs's use of this term as an indication that it does. But we can do more than this.

For the notion of prototype employed in the context of top-down pattern recognition to map onto the notion of a gestalt, a gestalt would need to be a complex entity consisting of in-principle separable elements that together form the prototype. It certainly sounds like a promising fit, if we remember that Parnas, Sass, and Zahavi (2013, p. 257) talk about how the "unity [of the gestalt] emerges from the relations between component features". However, we can back up this link even more strongly if we consider the notion of the gestalt from other sources. Ehrenfels, one of the founders of gestalt psychology, tells us that having a gestalt representation of something means having a

> content of presentation bound up in consciousness with the presence of complexes of mutually separable (i.e., independently presentable) elements. That complex of presentations which is necessary for the existence of a given *Gestalt* quality we call the foundation of that quality. (Ehrenfels 1890, in Smith, 1988, p. 93)

The gestalt (and then also the clinical gestalt of a disorder), understood in this way, is a whole consisting of related elements. Thus, understanding the phenomenological proposal along the lines of a prototype-based top-bottom pattern recognition process appears plausible.

As a result, the following picture emerges. According to Fuchs, the psychiatrist directly perceives the disorder as complex or gestalt in the patient after being confronted with diagnostic information and without further explicit cognitive efforts. This interpretation of Fuchs's general idea also matches with Parnas and colleagues' elaborations. As we saw, they claim (in the first quotation I offered above) that while the disorder and its symptoms are *ontologically* mutually constitutive, the clinical gestalt nevertheless enjoys *epistemic* primacy. As we saw in their last quote, they seem to believe that signs and symptoms are epistemically secondary insofar as the psychiatrist determines them after identifying the disorder. This interpretation follows from their statement that only once the gestalt of the disorder is recognised can a psychiatrist proceed to identify the symptoms and signs of that disorder in the patient. The gestalt must be recognised first, since only the clinical gestalt of the psychopathology allows for a symptom or sign to "articulate itself" and therefore become epistemically accessible to the psychiatrist. Reinterpreting this idea considering research in cognitive science, we may say that the phenomenological proposal for understanding psychiatric diagnostic reasoning is an automated form of prototype-based pattern recognition. This form of prototype-based pattern recognition leads psychiatrists to form cognitively unmediated assumptions (i.e., assumptions without explicit inferential reasoning) about the presence of a disorder in a patient that occurs to the psychiatrist as a gestalt quality of their perception of this patient. Any details of the psychopathological state of the patient are thereby epistemically secondary.[7] Particular features of a disorder can be accessible and become relevant

7 To interpret the phenomenologists' proposal as the idea that psychiatrists can directly access patients' overall psychopathological mental condition via a quasi-perceptual process fits with other views held by authors from the phenomenological tradition. Zahavi (2019), for example, defends a similar position, not regarding psychopathological mental conditions but for our overall interpersonal access. In his view, our everyday knowledge about each other's minds (e.g., about whether someone is angry) is also acquired in a direct quasi-perceptual manner without cognitive mediating processes.

to the psychiatrist only if the disorder is already recognised. The phenomenologists' proposal therefore has the feature of being epistemically top-down. It is a *disorder first, symptoms second* account of psychiatric diagnostic reasoning. Now that I have presented the phenomenological proposal, let me come to discuss it.

If we understand the phenomenological proposal as attempting to providing an adequate description of the method that underlies clinical psychiatric diagnostics, the first problem occurring is descriptive adequacy. This problem occurs in two forms. First, it results from the use of a very abstract, almost nonexistent, description of the diagnostic process itself onto which the phenomenologists map their proposal. The second version of the problem results from implausibilities concerning the proposed method itself. Let me discuss both in turn, starting with the aspect of the descriptive adequacy problem arising from the abstract basic picture of psychiatric diagnostics.

The phenomenological proposal provides only a rather abstract picture of what it takes to be the diagnostic process it is mapping onto. There is no mentioning of the steps of the diagnostic process, no talk about the screening taking place at the start of diagnostics, and no mention of the case formulation. It appears that either they presuppose a basic description of the process of psychiatric diagnostics that is so abstract (i.e., psychiatrists see patients and then diagnose them) that the interesting details of the diagnostic process are not explicitly discussed to any degree of detail, or they consider that good psychiatric diagnostics does not follow an approach that is more structured than this, for example as presented in my first chapter. In the first case, it seems that they assume an extremely abstract picture of psychiatric diagnostics, which, if we explained it by proposing a method that comprehensively covered it, would nonetheless offer only a very vague understanding of the actual diagnostic process in all its details – a rather unsatisfying result. If the second option is the case, then they simply seem to have an idiosyncratic understanding of psychiatric diagnostics, which also would disqualify their approach as relevant to understanding how what we would consider proper contemporary psychiatric diagnostics works. This would render their proposal uninteresting for the scope of this investigation, but also not opposed to my view. The more charitable interpretation, which also keeps the phenomenologists in the game as proposing an alternative to my account, would be to assume that they do wish to address what is considered proper psychiatric diagnostics, and not some rather totally different way of diagnosing patients. We should therefore interpret them as intending to adhere to professional standards rather than as considering a form of diagnosis that violates professional standards. If we do so, and thus assume the first case, their proposal nonetheless seems to be an unsatisfyingly abstract way to present a method of psychiatric diagnostics, due to their mostly nonexistent description of the process of diagnosing itself and the fact that, drawing on my discussion above, it appears that the phenomenological proposal makes no effort to be in touch with what is commonly assumed good contemporary psychiatric diagnostic practice. In its presentation, the phenomenological account lacks the connection to clinical reality, and thus seem unfit to be considered the presentation of a method that maps satisfyingly onto the aspects of what psychiatric clinicians do. But as mentioned, this is not the only point to consider. Even if we were satisfied with the vague picture of the actual diagnostic process that they operate with, there would be a problem with their proposal itself.

The second problem concerns the inconsistency of the phenomenological account with widely regarded empirical research on clinical diagnostics. In research conducted by psychologists and medical education researchers, two types of cognitive processes have been identified as relevant in diagnostic reasoning: bottom-up pattern recognition (e.g., Conderre et al., 2003; Groves, O'Rourke, and Alexander, 2003) and analytic reasoning (e.g., Croskerry, 2009). Since these two types of reasoning are widely recognised as being involved in diagnostic reasoning, any theory of diagnostic reasoning should either be coherent with the assumption that they are present, or if not, provide good reasons why – going against common sense and research – this is not the case. However, neither cognitive process has a place in the phenomenological account, nor does this account provide reasons why not to expect the presence of this type of reasoning. Let me elaborate.

At first glance, one might be inclined to interpret the phenomenological approach to gestalt recognition along the lines of bottom-up pattern recognition. However, it is not understood as such, at least in the context of research on medical cognition. In this context, bottom-up pattern recognition is considered a highly automatic, cue-based, feature-outcome associating process, whereas the cues are the signs and symptoms of the disorder, while the pattern is identified with the syndromal disorder diagnosis (Loveday et al., 2013). This, however, is not what is suggested in the phenomenological proposal as worked out earlier. While in bottom-up pattern recognition, symptoms and signs must be individuated and identified first, and only based on them is there an automatic detection of the disorder, the phenomenological proposal turns this process upside down. The phenomenological proposal, as a *disorder first, symptom second* approach, grants epistemic primacy to the disorder gestalt (i.e., the pattern). In their approach, the disorder must be recognised prior to the discerning of symptoms. It therefore appears that pattern recognition in the sense typically espoused by researchers in not included in the phenomenological proposal.

Analytic reasoning also plays no role in the phenomenological account. Analytic reasoning involves the explicit and careful consideration of the patient's presentation, identifying symptoms given certain background knowledge, and carefully weighing which diagnostic options are most plausible based on the available evidence. While pattern recognition is often used in simple diagnostic tasks (e.g., diagnosing a flu), analytic reasoning is commonly employed when medical experts face complex or ambiguous diagnostic scenarios (Croskerry, 2009). Such complexities and ambiguities often appear in psychiatric cases. On the phenomenological account, by contrast, the disorder is first recognised as a whole gestalt, and symptoms are individuated only after the gestalt of the disorder has been recognised in the patient. It therefore seems that analytic reasoning plays no part in the actual diagnostic reasoning process that identifies a disorder. If such reasoning is exercised at all, it would provide only a circular form of post-hoc justification for the diagnostic intuitions by which the clinician recognised the patient's disorder in the first place, since it is this initial diagnosis that forms the basis on which (rather than on any independent grounds) the confirming symptoms would be recognised. As the phenomenologists claimed in their quotes symptoms are epistemically individuated only in the context of the previously recognised gestalt. If psychiatrists really diagnosed in the manner described by the phenomenological account, it seems that they would not engage in analytic diagnostic reasoning.

Neither type of reasoning occurs in the context of the phenomenological proposal, and no reason is provided to explain why they should not occur. It therefore seems that the phenomenological proposal goes against what we should expect to be present in the context of psychiatric diagnostic reasoning, without any reasons that could support the rejection of the expectation that a proposal should cohere with research insights into the usual presence of pattern recognition and analytic reasoning in diagnostic reasoning. If the burden of making such an argument is not met, this seems to be a problem for the proposed method of the phenomenological account; its proposed method simply does not seem to be in line with what we should expect from a method of diagnostic reasoning. Hence, the description for their proposed method of diagnostic reasoning seems to be inadequate. It is inadequate because it is too abstract to qualify as a satisfyingly detailed understanding mapping onto the actual steps of psychiatric diagnostics, and also because in itself because it seems to be inconsistent with some well-founded expectations we can hold regarding a proposed method.

While the previously discussed point would apply to the phenomenological proposal no matter whether it intended to be a comprehensive answer to the Methodological Question or only an aspect of what would provide such an answer, there are additional problems if we assume for a moment that the former is true. If it wanted to present a full answer to the Methodological Question, the phenomenological proposal would fail to address two adequacy conditions for such an answer. First, it would not provide us with any rationale for the method they propose. There is no discussion of the rationale, the inferential strategy, or the inferential patterns at work in the *disorder first, symptom second* gestalt approach that would support its procedure, and this kind of discussion is needed for a methodology of the proposed method. Moreover, the topic of justification remains unaddressed. Due to the aforementioned lack of a rationale presented to back up their method, they cannot spell out the internal justification of their method – that is, what the method's internal principles are that ensure it justifies drawing a conclusion accruing to the method. Nor did they make plausible how the employment of a disorder gestalt could be supported *qua* external justification, because they provided no information on where these disorder gestalts used by psychiatrists come from and why they should be considered valid guides for diagnostics. In sum, it seems that the phenomenological proposal, wither we take it to be a full or only a partial attempt to address the Methodological Question, ends up being either an unsatisfying or an unsatisfying and incomplete proposal, respectively. Next, as indicated, I want to look at the problems the phenomenological proposal produces if we consider the desiderata for an answer to the Methodological Question.

There are several desiderata that the method proposed in the phenomenological proposal seem to be incapable of addressing. First, it does not address the intra- and interpersonal critical diagnostic reasoning that leads to revising one's own earlier diagnostic conclusions as well as critically discussing diagnostic conclusions among colleagues and rationally resolving disagreements. Second, the phenomenological proposal seems unable to identify diagnostic malpractice or to support its differentiation from mere misdiagnosis.

Critical diagnostic reasoning, as already discussed in the previous chapter, is used by clinical professionals who are trained in it and expected to practice it, no matter their specialisation (e.g., Marmaden, Schmidt, and Riekers, 2007; Harjai and Tiwari, 2009). Engaging in critical diagnostic reasoning means critically examining one's own or another's diagnostic judgements in order to avoid making mistakes in diagnostics due to biases or other errors in reasoning. Questions like "Why exactly should I draw this diagnostic conclusion?", "What could be an alternative explanation?", "Did I consider all available and potentially relevant information?" are typically asked when engaging in this kind of reasoning. Critical diagnostic reasoning can take place intrapersonally (by critically evaluating one's own diagnostic judgements) or interpersonally (by evaluating the diagnoses of others, as a clinician who supervises or works on a team might). To engage in critical reasoning about the justification of one's diagnosis in a non-circular way, however, analytical diagnostic reasoning is a prerequisite.

The problem here for the phenomenological account is that if a psychiatrist were to diagnose in the manner it prescribes, this intra- and interpersonal critical diagnostic reasoning would be impossible, or at least unnecessary. Intrapersonal critical reasoning would not be required, since considerations of a more plausible alternative diagnosis, given the symptoms and signs of disease, could not be found. In the phenomenological proposal, it is the initial diagnosis that determines what signs and symptoms the clinician will be able to individuate in the patient. It follows that any attempt to evaluate one's own diagnosis will, by the logic of the phenomenological account, lead to a necessarily self-confirming result. To get out of this vicious circle, the identification of signs and symptoms has to be achieved analytically as described above.

There is a related problem when it comes to interpersonal critical evaluation. Part of the critical interpersonal discussion of diagnoses involves explaining to other clinical experts why one has given a particular diagnosis. These experts suggest potential alternatives in order that they may eventually come to an agreement on the best diagnostic decision. If, however, all symptoms and signs that the clinicians recognise depend on their initial diagnoses, then pointing out other symptoms or signs to them would be hopeless because they would not be able to individuate those symptoms or signs independently of their original diagnosis. Interpersonal disagreements about diagnoses would become unresolvable and farcical, since any one participant could never rationally convince the other participants who disagreed with him. This is because it would be impossible for both sides to recognise the symptoms and signs that could serve as counterevidence to their own diagnostic proposal, given that those symptoms and signs would not fit the gestalt they recognised. The symptoms and signs individuated by each side in the disagreement would, at least in principle, be epistemically inaccessible to the other. It appears that in sum, the phenomenological proposal not only fails to provide an explanation for the intra- and interpersonal correction of diagnostic judgements, but moreover it is set up in a way that arguably makes it impossible for critical diagnostic reasoning – which we usually see and expect in the context of clinical diagnostics – to take place. Next up is malpractice.

To identify malpractice and distinguish it from mere misdiagnosis, what we need to be able to do is to identify what went wrong in the diagnostic process. We need to decide whether the wrong diagnosis was given due to missing, insufficient, or wrong information available to the clinician by reasonable information-gathering efforts, or whether the clinician themselves has done something wrong with the in-principle sufficient information base in the context of their diagnostic reasoning efforts. In the first instance, we would have a case of mere misdiagnosis; in the latter, it would be a case of malpractice. However, if we look at the phenomenological proposal, it is not clear how we should make this distinction.

We do not know what information is supposedly crucial for diagnostic decision-making according to the phenomenological approach, so we cannot evaluate when sufficient or insufficient information was attained and whether this information (or lack of it) should be considered responsible for a wrong diagnosis. Moreover, since the phenomenological approach provides us with no guidance on how the disorder gestalt is discovered by the psychiatrist, we have no way to assess whether, in the process of coming up with one's diagnostic conclusion *qua* the recognition of the supposed disorder gestalt, any mistake has taken place. Diagnostics according to the phenomenological approach remains a black box regarding the relevance of different types and tokens of information in any given diagnostic process and regarding the very process by which the diagnostic conclusions are drawn. The immunity towards being an object of meaningful critical diagnostic reasoning, as discussed in the previous paragraphs, therefore has the secondary effect that any result achieved by the recognition of the disorder gestalt also seems to be unfit to be evaluated as potentially being a case of misdiagnosis or malpractice if it turns out to be wrong.

Having pointed out these problems of the phenomenological account, let me now compare how the model-based proposal holds up against it in all these problem domains. First, regarding descriptive adequacy, it seems that the model-based approach performs better than the phenomenological approach. As briefly discussed at the start of the last chapter, my proposal meets the adequacy requirement of being cognitively realistic, which entails that its proposed explanation of psychiatric diagnostic reasoning is true to a detailed description of the steps of the diagnostic procedure and ensures that the assumed inferential steps make sense of these procedures. Second, my discussion of how the model-based proposal explains intra- and interpersonal critical diagnostic reasoning also shows that it is able to put forward a plausible and helpful proposal on this front, which, as discussed, is fully blocked for the phenomenological proposal, which even undermines the possibility of such reasoning taking place. Thirdly and finally, regarding the inclusion of the empirically supported types of reasoning that are commonly encountered in the context of diagnostic reasoning, but that do not seem to play a role in the diagnostic proposal of the phenomenologists, again the model-based account holds up well. As I proposed in Chapter 3 and repeated in Chapter 4, the inference from symptoms to disorders can potentially (in well-trained diagnostic experts) be conceptualised as a rule-based pattern recognition process. Moreover, analytic reasoning plays a prominent role in the model-based account. This account assumes that the decision as to which complaint should be evaluated as constituting which kind of psychiatric or medical symptom is a detailed and thorough process that is carried out in the context of diagnostic reasoning, and then again when it is explicated in the context of the required case formulation that puts together the diagnostic conclusions at the level of symptoms as well as the disorder level, and supports them by the information considered crucial to support the diagnostic conclusions thus drawn.

In sum, the phenomenological proposal has at least two significant problems. The first major problem is its detachment from actual clinical diagnostic practices. Its proposed method seems to be inconsistent with plausibly expected features of a method of psychiatric diagnostic reasoning. The second problem is its inability to explain critical diagnostic reasoning and to help us to understand and discern the differences between diagnostic mistakes and malpractice. If we evaluate it as a full-blown proposal to address the Methodological Question, we would have to add that it does not address two of the relevant adequacy conditions – namely, providing the rationale for the method's operations and demonstrating how we should consider the results of the method to be justified, both internally and externally. All these are points in which the model-based account I have presented over the preceding chapters performs better. It is intimately close to actual diagnostic practice, as shown in Chapters 1 and 3; it can make sense of intra- and interpersonal critique and revisions of diagnostic decisions as we see them every day in the clinic, as shown in Chapter 4, and it can help us to understand the difference between malpractice and misdiagnosis and provides guidance on how to assess which of the two took place, as also discussed in Chapter 4. Considering these problems of the phenomenological approach that the model-based approach does not encounter, it seems the model-based proposal is preferable.

5.6 Conclusion

In this chapter I selected several philosophical contributions to the topic of psychiatric diagnostics that *prima facie* presented alternatives to or problems for my account to psychiatric diagnostic reasoning. I showed that on closer inspection these accounts variously turn out to not actually not concern psychiatric diagnostics itself (Cooper); to concern it, but in a way that is in principle compatible with my approach, though my approach does a better job of providing detailed discussions of the overall process and how to use this understanding to address relevant topics in the context of psychiatric diagnostics (Reznek and Murphy), or to concern aspects of psychiatric diagnostics that my proposal also deals with in a way that seem incompatible with my own approach, but when putting forward criticism or alternatives to my account, to be plagued by problems that make their proposals less plausible than my model-based account (Gupta, Potter, and Goyer).Finally I discussed the phenomenological proposal (Fuch, Sass, Parnas, Zahavi) which I showed to fail several adequacy conditions for a proper answer to the Methodological Question and to be detached from the clinical reality of diagnostics, making the model-based account I defend preferable over it. In the end, it seems that my account is the best candidate – one that, as demonstrated in the previous chapter, meets all adequacy conditions, allows us to address several interesting sub-questions regarding psychiatric diagnostics, and does so better than any of the candidates discussed in this chapter. The model-based account of psychiatric diagnostics seems to be the most well-rounded candidate to provide an answer to the Methodological Question.

7. Conclusion

How does a psychiatrist know whether a patient is mentally ill and, if so, what their specific condition is? This question was the starting point of my inquiry. Refining this into what I called the Methodological Question, I considered what the method of proper, contemporary clinical diagnostic reasoning may be – a question that must be answered by philosophy of psychiatry if what we aim for is a systematic understanding of the various aspects of the epistemology of psychiatry. But beyond this interest for the sake of knowledge itself, answering the Methodological Question also has practical implications justifying its pursuit. These include motivations from ethics and matters of law, since only a sufficiently general understanding of what method should be pursued in contexts of diagnostics allows us to evaluate whether the diagnostic work of clinicians violates the standards of the proclaimed method. If we have enough information about the case, this kind of general understanding puts us in a position to determine whether harm caused to a patient due to a false diagnosis is the result of malpractice or cannot be blamed on the diagnostician. Also, only if we understand how diagnostic judgements are formed and justified can we make case-by-case decisions about situations in which patients' judgements about their mental conditions differ from clinicians' judgements, to evaluate whether there are better epistemic reasons to believe the psychiatrist or the patient. Both issues are highly relevant. They are too complex to have been explored in this thesis, but the groundwork for potential future discussion of such issues has now been laid.

Another more pragmatic implication that makes an answer to the Methodological Question desirable is its potential value for medical education. The model developed here has the potential to be used in the theoretical and practical training of psychiatrists. Not that the answer I have proposed has provided any medical details that would be of value for this, but it has provided an abstract description for a method and its steps that in any case of diagnostic practice might be a background algorithm that could be taught in medical education. This would mean that a clinician could, by working though the steps of the method, assess for themselves whether the concrete steps of the diagnostic work can be subsumed under the more general methodological framework I proposed. Also in medical education, concrete cases could, for educational purposes, be discussed along the lines of this model for psychiatric diagnostics, to show individuals pursuing medical education what kinds of principles and structures stand behind the confusing first impressions that a clinician new to clinician practice may have of the overall process.

A final proactive application for this work that we may envisage is its usefulness for research. One application area might be attempts to build automatised diagnostic programs working with patient data, or tools that are supposed to support clinicians in aspects of their diagnostic reasoning process; both could be modelled following the method proposed here. In the case of fully automated systems, for example, this would ensure that the systems operated according to the same standards, and by going through the same steps, that we expect to be adhered to by clinicians under ideal practical circumstances, which in turn may help us develop understandable and more trustworthy fully automated solutions for psychiatric diagnostics. Or, if we were simply aiming to develop tools to support diagnostic work, we could, for example, model systems that support differential diagnostic processes by providing propositional models to psychiatrists if they enter a complaint, helping them to ensure that they did not forget a potential evaluation, and providing them with a good user experience because the tool presents information in a format that is close to their own cognitive efforts. As we see, there are many reasons to try to answer the Methodological Question.

The answer to the Methodological Question I have presented in this thesis is the model-based account of diagnostic reasoning. It is intended to present a methodology providing us with a description of what can be understood to be the method behind the belief-forming procedures in psychiatric diagnostics, and explaining what the rationale behind the operations of these methods are and how this method is supposed to ensure that its results are justified. By being intended to meet these constraints, it should meet what I considered to be the adequacy conditions for an answer to the Methodological Question. In addition to this, in presenting my proposal I have aimed to provide a framework that would allow us to address psychiatric diagnostics in a way that is especially satisfactory regarding how it addresses diagnostic reasoning and several relevant phenomena in this context. I called these extra things I wanted from a proposal desiderata, and they are that the proposed answer should be comprehensive, cognitively realistic, helpful for making sense of the difference between misdiagnosis and diagnostic malpractice, accounting for the occurrence and resolution of diagnostic uncertainty, helpful for understanding and evaluating the phenomenon of good instinctual diagnostics and the occurrence and solution of diagnostic disagreements, and finally showing the right degree of robustness as well as falsifiability in relation to changes in psychiatric science and diagnostic practice.

The model-based account of diagnostic reasoning was developed to meet all these requirements. To sum up, the idea is that psychiatric diagnostics should be understood as a qualitative, constitutive diagnostic modelling process. To establish this idea, I began by providing a conmonnsensical description of the constitutive core aspects of psychiatric diagnostics based on the psychiatric training and guild line literature representing the standard core procedures of proper contemporary diagnostic reasoning. I then spelled out the methodology that I intended to apply to the process of modelling thus understood. I explained what makes something modelling, described when modelling is qualitative and constitutive, and introduced the specific modelling framework of diagnostic modelling. Subsequently, I came back to the process of psychiatric diagnostics, this time looking at more particular clinical instances, and mapped out step by step the various features of the previously proposed methodology of the clinical diagnostic process, showing that the method of modelling I proposed, and accordingly the other aspects of the methodology behind it, seem to adequately apply to clinical psychiatric diagnostics. After providing this adequate methodology, I let my answer to the Methodological Question do some heavy lifting. I showed how each of the proposed desiderata is fulfilled by my account, making the model-based approach a satisfyingly adequate and indeed particularly fruitful answer to the Methodological Question. Since my own proposal is not the only game in town, I turned towards supposed alternatives to my account and potential criticisms that would apply to it. I evaluated each of the alternative proposals and responded to all the critical accounts under consideration, concluding that the model-based account is the most satisfying of all the proposals, and that the discussed criticism does no relevant harm to my proposal.

By introducing and defending the model-based account as the first systematic and fully mapped out approach to applying debates about modelling in philosophy of science to the topic of medical diagnostics, more specifically to psychiatric diagnostics, I hope to have made a stimulating contribution to debates in the epistemology of psychiatric diagnostics, a still small aspect of the overall debates in the field. I also hope to have provided an example of how debates from general philosophy of science, in this case about modelling, can be made fruitful in the application to discussions of aspects of special sciences and practices such as psychiatry. Finally, if anything I have done in these pages impacts any of the pragmatic concerns of psychiatry that I mentioned above – if it ever proves useful in medical education, inspires debates about policies on how to differentiate misdiagnosis, or inspires a developer in health tech to come up with a useful program or device, and indeed if it ever leads to something that that is of help to anyone seeking psychiatric treatment – I will consider this research to have served its purpose. Future work in these areas, through which I hope to build on this thesis, will help to make this hope a reality.

Sources

Aarseth, E., Bean, A.M., Boonen, H., Colder Carras, M., Coulson, M., Das, D., Deleuze, J., Dunkels, E., Edman, J., Ferguson, C.J., Haagsma, M.C., Helmersson Bergmark, K., Hussain, Z., Jansz, J., Kardefelt-Winther, D., Kutner, L., Markey, P., Nielsen, R.K.L., Prause, N., Przybylski, A., Quandt, T., Schimmenti, A., Starcevic, V., Stutman, G., Van Looy, J., and Van Rooij, A.J. (2017) 'Scholars' open debate paper on the World Health Organization ICD-11 Gaming Disorder proposal', *Journal of Behavioral Addiction*, 6(3), pp. 267–270. doi: 10.1556/2006.5.2016.088

Abbott K.H., Schutte P.C., Palmer M.T., Ricks W.R. (1987) '*Faultfinder: A diagnostic expert system with graceful degradation for onboard aircraft applications*', Paper presented at 14th International Symposium on Aircraft Integration Monitoring Systems. Friedrichshafen, West Germany.

Aboraya, A. (2009) 'Use of structured interviews by psychiatrists in real clinical settings: Results of an open-question survey', *Psychiatry*, 6(6), pp. 24–28.

Abou Taam, M., Boissieu, P.D., Abou Taam, R., Breton, A., and Trenque, T. (2015) 'Drug-induced hallucination: a case/non case study in the French Pharmacovigilance Database', *The European Journal of Psychiatry*, 29 (1), pp. 21–31. doi: 10.4321/S0213-61632015000100002

Adam, D. (2013) 'Mental health: On the spectrum', *Nature*, 496, pp. 416–418. doi: 10.1038/496416a

Ægisdóttir, S., White, M.J., Spengler, P.M., Maugherman, A.S., Anderson, L.A., Cook, R.S., and Rush, J.D. (2006), 'The meta-analysis of clinical judgment project: Fifty-six years of accumulated research on clinical versus statistical prediction', *The Counseling Psychologist*, 34(3), pp. 341–382. doi: 10.1177/0011000005285875

Akeret, R.U. (1995) *The Man who Loved a Polar Bear: And Other Psychotherapist's Tales.* London: Penguin Books.

Alexandrova, A. (2017) *A Philosophy for the Science of Well-being.* Oxford: Oxford University Press.

Aliseda, A. (2006) *Abductive Reasoning*. Dordrecht: Springer.

Altable, C.R. (2012) 'Logic structure of clinical judgment and its relation to medical and psychiatric semiology', *Psychopathology*, 45(6), pp. 344–351. doi: 10.1159/000337968

American Board of Psychiatry & Neurology. (2019) *Homepage of the American Board of Psychiatry & Neurology: Training Infomration*. Available at:https://www.abpn.com/access-residency-info/residency-training-information/psychiatry/ (Accessed 1 January 2019).

American Psychiatric Association (1980) *Diagnostic and Statistical Manual of Mental Disorders*. 3rd edn. Arlington, VA: American Psychiatric Association.

American Psychiatric Association (2013) *Diagnostic and Statistical Manual of Mental Disorders*. 5th edn. Washington, DC: American Psychiatric Association.

American Psychiatric Association (2022) *Diagnostic and Statistical Manual of Mental Disorders*. 5th end. Text Revision. Washington, DC: American Psychiatric Association.

Andreasen, N.C. (2016) 'Thought disorder', in Fatemi, S.H. and Clayton P. J. (ed.), *The medical basis of psychiatry*. 4th edn. pp. New York: Springer Science+Business Media, pp. 497–505.

Angner, E. (2011) 'Are subjective measures of well-being "direct"?', *Australasian Journal of Philosophy*, 89(1), pp. 115–130. doi: 10.1080/00048400903401665

Anselme, P. and Robinson, M.J.F. (2019) 'Incentive motivation: The missing piece between learning and behavior', in Renninger, K.A. and Hidi S. E. (ed.) *The Cambridge Handbook of Motivation and Learning*, Cambridge: Cambridge University Press, pp. 163–182.

Aliseda, A. and Leonides, L. (2013) 'Hypotheses testing in adaptive logics: an application to medical diagnosis', *Logic Journal of the IGPL*, 21, pp. 915–930. doi: 10.1093/jigpal/jzt005

Attarian, B., Raizer, J., Kumthekar, P., Macken, M.P., Schuele, S.U., and Gerard, E. (2013) 'Auditory hallucinations as ictal phenomena in a patient with voltage-gated potassium channel antibody-associated limbic encephalitis', *Epileptic Disorders*, 15 (4), pp. 433–436. doi: 10.1684/epd.2013.0623

Audétat, M.C., Dory, V., Nendaz, M., Vanpee, D., Pestiaux, D., Junod Perron, N., and Charlin, B. (2012) 'What is so difficult about managing clinical reasoning difficulties?', *Medical Education*, 46 (2), pp. 216–227. doi: 10.1111/j.1365-2923.2011.04151.x

Bach, B., First, M.B. (2018) 'Application of the ICD-11 classification of personality disorders', *BMC Psychiatry* 18, pp. 351–365. doi: 10.1186/s12888-018-1908-3

Bajpai, S.R. and Bajpai, R.C. (2014) 'Goodness of Measurement: Reliability and Validity', *International Journal of Medical Science and Public Health*, 3(2), pp. 112–115. doi: 10.5455/ijmsph.2013.191120133

Baker, M.G., Kale, R. and Menken, M. (2002) 'The wall between neurology and psychiatry: Advances in neuroscience indicate it's time to tear it down. ' *BMJ*, 324, pp.1468. doi: 10.1136/bmj.324.7352.1468

Barbano, A.C., van der Mei, W.F., Bryant, R.A., Delahanty, D.L., deRoon-Cassini, T.A., Matsuoka, Y.J., Olff, M., Qi, W., Ratanatharathorn, A., Schnyder, U., Seedat, S., Kessler, R.C., Koenen, K.C. and Shalev, A.Y. (2019) 'Clinical implications of the proposed ICD-11 PTSD diagnostic criteria', *Psychological Medicine*, 49(3), pp.483-490. doi: 10.1017/S0033291718001101

Batterman, R.W. (2002) 'Asymptotics and the role of minimal models', *British Journal for the Philosophy of Science*, 53, pp. 21–38.

Beck, A.T. (1987) 'Cognitive models of depression', *Journal of Cognitive Psychotherapy: An International Quarterly*, 1, pp. 5–37.

Beck, A.T. (1967) *Depression: Clinical, Experimental and Theoretical Aspects*. New York: Harper & Row.

Beck, A.T., Ward, C.H., Mendelson, M., Mock, J. and Erbaugh, J. (1961) 'An inventory for measuring depression', *Archives of General Psychiatry*, 4(6), pp. 561–571. doi: 10.1001/archpsyc.1961.01710120031004

Bell, D., Bobrow, D., Falkenhainer, B., Fromherz, M., Saraswat, V. and Shirley, V. (1994) 'RAPPER: The Copier Modelling Project', *In Proceedings of QR94*, pp.1-10.

Berridge, K.C. and Robinson, T.E. (1998) 'What is the role of dopamine in reward: hedonic impact, reward learning, or incentive salience?', *Brain Research Reviews*, 28, pp. 309–369. doi: 10.1016/s0165-0173(98)00019-8

Berrios, G.E. and Olivares, J.M. (1995) 'The anhedonias: a conceptual history', *History of Psychiatry*, 6(24), pp. 453–470. doi: 10.1177/0957154X9500602403

Bishop, M.A. and Trout, J.D. (2005) *Epistemology and the Psychology of Human Judgment*. Oxford: Oxford University Press.

Blalock, H.M. (1964). *Causal Inference in Nonexperimental Research*. Chapel Hill: University of North Carolina Press.

Bobrow, D.G. (1984). 'Qualitative reasoning about physical systems: an introduction', *Artificial Intelligence*, 24 (1), pp. 1–5. doi: 10.1016/0004-3702(84)90036-5

Bogacz, R. (2020) Dopamine role in learning and action inference. *eLife*, 9, e53262. doi: 10.7554/eLife.53262

Boland, R., Verdiun, M. and Ruiz, P. (2021) *Kaplan & Sadock's Synopsis of Psychiatry 12th Edition*. Philadelphia: Lippincott Williams & Wilkins.

Bollen K.A., Davis WR. (2009) 'Causal indicator models: Identification, estimation, and testing', *Structural Equation Modeling*, 16 (3), pp. 498–522. doi: 10.1080/10705510903008253

Bollen, K.A. and Bauldry, S. (2011) 'Three Cs in measurement models: causal indicators, composite indicators, and covariates', *Psychological Methods*, 16 (3), pp. 265–284. doi: 10.1037/a0024448

Boorse, C. (1977). Health as a theoretical concept. *Philosophy of Science*, 44(4), pp. 542–573. doi: 10.1086/288768

Borkovec, T.D., Robinson, E., Pruzinsky, T. and DePree, J.A. (1983) 'Preliminary exploration of worry: some characteristics and processes', *Behaviour Research and Therapy*, 21(1), pp. 9–16. doi: 10.1016/0005-7967(83)90121-3

Borkovec, T.D. and Inz, J. (1990) 'The nature of worry in generalized anxiety disorder: a predominance of thought activity', *Behaviour Research and Therapy*, 28, pp.153-158. doi: 10.1016/0005-7967(90)90027-g

Borsboom, D. (2005) *Measuring the Mind: Conceptual Issues in Contemporary Psychometrics.* Cambridge: Cambridge University Press.

Borsboom, D. (2017). 'A network theory of mental disorders', *World Psychiatry*, 16(1), pp. 5–13. doi: 10.1002/wps.20375

Borsboom, D., Mellenbergh, G., Van Heerden, J. (2004) 'The concept of validity', *Psychological Review*,111 (4), pp. 1061–1071. doi: 10.1037/0033-295X.111.4.1061

Borsini, A., Wallis, A.S.J., Zunszain, P., Pariante, C.M., and Kempton, M.J. (2020) 'Characterizing anhedonia: A systematic review of neuroimaging across the subtypes of reward processing deficits in depression', *Cognition and Affective Behavior Neuroscience*, 20 (4), pp. 816–841. doi: 10.3758/s13415-020-00804-6

Boumans, M. (1999) 'Built-in justification', in Morgan, M. S. and Morrison, M. (ed.) *Models as Mediators: Perspectives on Natural and Social Science*. Cambridge: Cambridge University Press, pp. 66–96.

Bruch, M. and Bond, F.W. (2015) *Beyond Diagnosis: Case Formulation in Cognitive Behavioural Therapy*. Chichester: John Wiley & Sons.

Bruchmüller, K., Margraf, J., Suppiger, A. and Schneider, S. (2011) 'Popular or unpopular? Therapists' use of structured interviews and their estimation of patient acceptance', *Behavior Therapy*, 42 (4), pp. 634–643. doi: 10.1016/j.beth.2011.02.003

Burge, T. (2010) *Origins of Objectivity*. Oxford: Oxford University Press.

Butler, G. (1998) 'Clinical formulation', in Bellack, A. S. and Hersen, M. (ed.), *Comprehensive clinical psychology*. Oxford: Pergamon, pp. 1–23.

Cahalan, S. (2019) *The great pretender: The undercover mission that changed our understanding of madness*. New York City: Hachette.

Carter, J.A. and Littlejohn, C. (2021) *This is Epistemology: An Introduction*. New York City: John Wiley & Sons.

Cartwright, N. (1983) *How the laws of physics lie.* Oxford: Oxford University Press.

Cartwright, N. (1989) *Nature's capacities and their measurement.* Oxford: Oxford University Press.

Cartwright, N. (2022) 'How to Learn about Causes in the Single Case', in Widner, J., Woolcock, M. and Ortega Nieto, D. (ed.), *The Case for Case Studies: Methods and Applications in International Development*. Cambridge: Cambridge University Press, pp. 29–51.

Cartwright, N., Bradburn, N. (2011) 'A Theory of Measurement', in National Research Council (ed.), *The Importance of Common Metrics for Advancing Social Science Theory and Research: Proceedings of the National Research Council Committee on Common Metrics*. Washington, DC: National Academies, pp. 53–70.

Carvalho, A.F., Solmi, M., Sanches, M., Machado, M.O., Brendo, S., Ajnakian, O., Sherman, C., Sun, Y.R., Liu, CS, Brunoni, A.R., Pigato, G., Fernandes, B.S., Bortolato, B. Muhammad, I.H., Dragioti, E., Firth, J. Cosco, T.D., Maes, M., Berk, M., Lanctôt, K.L., Vieta, E., Pizzagalli, D.A., Smith, L., Fusar-Poli, P., Kurdayka, P.A., Fornaro, M., Rehm, J., and Hermann, N. (2020) 'Evidence-based umbrella review of 162 peripheral biomarkers for major mental disorders', *Translational Psychiatry*, 10, pp. 152–165. doi: 10.1038/s41398-020-0835-5

Casey, P. and Kelly, B. (2019) *Fish's Clinical Psychopathology: Signs and Symptoms in Psychiatry. 4th Edition*. Cambridge: Cambridge University Press.

Choong, C., Hunter, M.D. and Woodruff, P.W. (2007) 'Auditory hallucinations in those populations that do not suffer from schizophrenia', *Current Psychiatry Reports*, 9(3), pp. 206–212. doi: 10.1007/s11920-007-0020-z

Churchland, P.M. (1981) 'Eliminative materialism and propositional attitudes', *The Journal of Philosophy*, 78(2), pp. 67–90. doi: 10.2307/2025900

Clark, D.A. and Claybourn, M. (1997) 'Process characteristics of worry and obsessive intrusive thoughts', *Behaviour Research and Therapy*, 35, pp. 1139–1141. doi: 10.1016/S0005-7967(97)80008-3

Clark, D.A. and Rhyno, S. (2005) 'Unwanted intrusive thoughts in nonclinical individuals', in Clark, D.A. (ed.) *Intrusive thoughts in clinical disorders: Theory, research, and treatment*. New York City: The Guilford Press, pp. 1–29.

Clark, D. A., Beck, A. T. and Alford, B. (1999) *Scientific Foundations of Cognitive Theory and Therapy of Depression*. New York: Wiley.

Coderre, S., Mandin, H., Harasym, P. H., and Fick, G.H. (2003) 'Diagnostic reasoning strategies and diagnostic success', *Medical education*, 37 (8), pp. 695–703. doi: 10.1046/j.1365-2923.2003.01577.x

Cooper, R. (2014) *Psychiatry and Philosophy of Science*. Abingdon: Routledge.

Covell, T. and Siddiqui, W. (2022) *Korsakoff Syndrome*. Available at: https://www.ncbi.nlm.nih.gov/books/NBK539854/ (Accessed 20 July 2022).

Craig, W. (1918) 'Appetites and aversions as constituents of instincts', *The Biological Bulletin*, 34(2), pp. 91–107. doi: 10.1073/pnas.3.12.685

Craver, C. (2007) 'Constitutive explanatory relevance'. *Journal of Philosophical Research*, 32, pp.3-20. doi: 10.5840/jpr20073241

Croskerry, P. (2009) 'A universal model of diagnostic reasoning', *Academic Medicine: Journal of the Association of American Medical Colleges*, 84(8), pp. 1022–1028. doi: 10.1097/ACM.0b013e3181ace703

Croxson, P.L., Walton, M.E., O'Reilly, J.X., Behrens, T.E. and Rushworth, M.F. (2009) 'Effort-based cost-benefit valuation and the human brain', *Journal of Neuroscience*, 29 (14), pp. 4531–4541. doi: 10.1523/JNEUROSCI.4515-08.2009

Cummins, R.C. (1975) 'Functional Analysis'. *The Journal of Philosophy*, 72, pp.741-765. doi: 10.2307/2024640

Cummins, R.C. (1983) *The Nature of Psychological Explanation*. Cambridge: Bradford/ MIT Press.

Cummins, R.C. (2000) '"How does it work" versus "what are the laws?": Two conceptions of psychological explanation', in Keil, F. and Wilson, R.A. (ed.), *Explanation and Cognition*. Cambridge: Bradford/MIT Press, pp. 117–145.

Cuthbert, B.N. and Insel, T.R. (2013) 'Toward the future of psychiatric diagnosis: the seven pillars of RDoC', *BMC Medicine*, 11, pp. 126–134. doi: 10.1186/1741-7015-11-126

Cuthbert, B.N. and Morris, S.E. (2021) 'Evolving concepts of the schizophrenia spectrum: A research domain criteria perspective', *Frontiers in Psychiatry*, 12, 641319. doi: 10.3389/fpsyt.2021.641319

Krystal, A.D., (2012) Psychiatric disorders and sleep. *Neurologic clinics*, 30 (4), pp.1389-1413. doi: 10.1016/j.ncl.2012.08.018

Da Costa, N. and French, S. (2003) *Science and Partial Truth: A Unitary Approach to Models and Scientific Reasoning.* New York, US: Oxford University Press US.

David, A.S. and Nicholson, T. (2015) 'Are neurological and psychiatric disorders different?', *The British Journal of Psychiatry*, 207 (5), pp.373-374. doi: 10.1192/bjp.bp.114.158550

Dayan, P. (2002) 'Reinforcement learning', in Pashler, H. and Gallistel, R. (ed.), *Steven's Handbook of Experimental Psychology: Learning, Motivation, and Emotion*. New York: John Wiley & Sons Inc., pp. 103–129.

De Fruyt, J., Sabbe, B. and Demyttenaere, K. (2020) Anhedonia in depressive disorder: a narrative review, *Psychopathology*, 53 (5), pp. 274–281. doi: 10.1159/000508773

de Kleer, J. and Brown, J.S. (1984). A qualitative physics based on confluences. *Artificial Intelligence*, 24, pp. 7–8. doi: 10.1016/0004-3702(84)90037-7

Dellazizzo, L., Potvin, S., Giguère, S., and Dumais, A. (2022) 'Evidence on the acute and residual neurocognitive effects of cannabis use in adolescents and adults: a systematic meta-review of meta-analyses', *Addiction* 117 (7), 1857–1870. doi: 10.1111/add.15764

Demazeux, S. and Singy, P. (ed.) (2015) *The DSM-5 in Perspective: Philosophical Reflections on the Psychiatric Babel*. Dordrecht: Springer.

Deng, H., Li, L.Y., Wang, Y., Wu, D., Gao, B., and Yang, X. (2012). 'Long-term follow-up of patients treated for psychotic symptoms that persist after stopping illicit drug use', *Shanghai Archives of Psychiatry*, 24 (5), 271–278. doi: 10.3969/j.issn.1002-0829.2012.05.004

Derrida, J., (1995) On the name. Redwood City: Stanford University Press.

Dichter, G.S., Felder, J.N., Petty, C., Bizzell, J., Ernst, M. and Smoski, M.J. (2009) 'The effects of psychotherapy on neural responses to rewards in major depression' *Biological Psychiatry*, 66 (9), pp. 886–897. doi: 10.1016/j.biopsych.2009.06.021

Division of Clinical Psychology (2010), *Core purpose and philosophy of the Profession. Leicester: The British Psychological Society*, available at: https://www.bps.org.uk/sites/default/files/documents/core_purpose_and_philosophy_of_the_profession.pdf (Accessed 15 April 2020).

Docherty, A.R., Berenbaum, H. and Kerns, J.G. (2011) 'Alogia and formal thought disorder: differential patterns of verbal fluency task performance', *Journal of Psychiatric Research*, 45 (10), pp. 1352–1357. doi: 10.1016/j.jpsychires.2011.04.004

Dougall, A.L., Craig, K.J. and Baum, A. (1999) 'Assessment of characteristics of intrusive thoughts and their impact on distress among victims of traumatic events', *Psychosomatic Medicine*, 61 (1), pp. 38–48. doi: 10.1097/00006842-199901000-00008

Doughty, O.J. and Done, D.J. (2009) 'Is semantic memory impaired in schizophrenia? A systematic review and meta-analysis of 91 studies', *Cognitive Neuropsychiatry*, 14 (6), pp. 473–509. doi: 10.1080/13546800903073291

Dretske, F. (1981). *Knowledge and the Flow of Information*. Cambridge: MIT Press.

Dunbar, R.I. (1992). 'Neocortex size as a constraint on group size in primates', *Journal of human evolution*, 22 (6), pp. 469–493. doi: 10.1016/0047-2484(92)90081-J

Durning, S.J., Artino, A.R. Jr., Beckman, T.J., Graner, J., van der Vleuten, C., Holmboe, E. and Schuwirth, L. (2013) 'Does the think-aloud protocol reflect thinking? Exploring functional neuroimaging differences with thinking (answering multiple choice questions) versus thinking aloud', *Medical Teacher*, 35 (9), pp. 720–726. doi: 10.3109/0142159X.2013.801938

Edmondson, A.J., Brennan, C.A., and House, A.O. (2016) 'Non-suicidal reasons for self-harm: A systematic review of self-reported accounts', *Journal of affective disorders*, 191, 109–117. doi: 10.1016/j.jad.2015.11.043

Eells, T. D. (2006) History and current status of psychotherapy case formulation, in Eells, T.D. (ed.) *Handbook of psychotherapy case formulation. 2nd Edition*. New York: The Guildford Press, pp. 3–32.

Egerton A., Mehta M.A., Montgomery A.J., Lappin, J.M., Howes, O.D., Reeves, S.J., Cunninham, and V.J., Grasby, P.M. (2009) 'The dopaminergic basis of human behaviors: A review of molecular imaging studies', *Neuroscience and Biobehavioral Reviews*, 33, pp. 1109–1132. doi: 10.1016/j.neubiorev.2009.05.005

Ehrenfels, C. [1890] (1980) 'On „Gestalt Qualities" ' in Smith, B. (ed.) *Foundations of Gestalt Theory*. München: Philosophia Verlag, pp. 82–155.

Elgin, C.Z. (2017) *True Enough*. Cambridge: MIT Press.

Ellingsen, D.M., Leknes, S. and Kringelbach, M.L. (2015) 'Hedonic value', in Gebauer, G.F., Schmaltz, J.A., and Slife, M.R. (ed.) *Handbook of Value: Perspectives from Eco-*

nomics, Neuroscience, Philosophy, Psychology and Sociology. Oxford: Oxford University Press, pp. 265–286.

Elstein, A.S., Shulman, L.S., and Sprafka, S.A. (1978) *Medical Problem Solving: An Analysis of Clinical Reasoning*. Cambridge: Harvard University Press.

Elstein, A.S. (1999) 'Heuristics and biases: selected errors in clinical reasoning', *Academic Medicine*, 74 (7), pp. 791–794. doi: 10.1097/00001888-199907000-00012

Elstein, A.S. (2009) 'Thinking about diagnostic thinking: A 30-year perspective', *Advances in Health Sciences Education*, 14 (1), pp. 7–18. doi: 10.1007/s10459-009-9184-0

Esterberg, M.L. and Compton, M.T. (2009) 'The psychosis continuum and categorical versus dimensional diagnostic approaches', *Current Psychiatry Reports*, 11, pp. 179–184. doi: 10.1007/s11920-009-0028-7

Falkenhainer, B. and Forbus, K.D. (1991) 'Compositional modelling: Finding the right model for the job', *Artificial Intelligence*, 51, pp. 95–143. doi: 10.1016/0004-3702(91)90109-W

Farley, A. M. (1987) 'Qualitative modelling of economic systems', in Ifac (ed.) *Economics and Artificial Intelligence: Proceedings of the Ifac/ifors/ifip/iasc/afcet Conference, Aix-En-provence, France, 2–4 September 1986*. Oxford: Pergamon, pp. 41–44.

Faucher, L. and Forest, D. (ed.) (2021) *Defining Mental Disorder: Jerome Wakefield and His Critics*. Cambridge: MIT Press.

Feigl, H. (1967) [1958] 'The 'Mental' and the 'Physical'', in: Feigl, H., Scriven, M. and Maxwell, G. (ed.) *Concepts, Theories and the Mind-Body Problem*. Minneapolis: Minnesota, pp. 370–497.

Feinstein, A.R. (1964) 'Scientific methodology in clinical medicine IV. Acquisition of clinical data', *Annals of Internal Medicine*, 61, pp. 1162–1193. doi:10.7326/0003-4819-61-6-1162

Ferretter, L. (2001) 'How to avoid speaking of the other: Derrida, Dionysius and the problematic of negative theology', *Paragraph*, 24 (1), pp. 50–65. doi: 10.3366/para.2001.24.1.50

Fiorentini, A., Cantù, F., Crisanti, C., Cereda, G., Oldani, L. and Brambilla, P. (2021) 'Substance-Induced Psychoses: An Updated Literature Review', *Front Psychiatry*, 12, 694863. doi: 10.3389/fpsyt.2021.694863

Forbus, K.D. (1984). 'Qualitative process theory'. *Artificial Intelligence*, 24, pp. 85–168.

Forbes, K.D. (2008), 'Qualitative modelling', in Van Harmelen, F., Lifschitz, V. and Porter, B. (ed.), *Handbook of knowledge representation*. Amsterdam: Elsevier, pp. 361–387.

Frigg, R. and Hartmann, S. (2020) 'Models in Science', The Stanford Encyclopedia of Philosophy (Spring 2020 Edition), ed. by Edward N. Zalta, available at: <https://plato.stanford.edu/archives/spr2020/entries/models-science/> [accessed 28 April 2020].

Friston, K.J., Stephan, K.E., Montague, R. and Dolan, R.J. (2014) 'Computational psychiatry: the brain as a phantastic organ', *The Lancet Psychiatry*, 1 (2), pp. 148–158. doi: 10.1016/S2215-0366(14)70275-5

Fryer, S.L., Roach, B.J., Holroyd, C.B., Paulus, M.P., Sargent, K., Boos, A. et al. (2021) 'Electrophysiological investigation of reward anticipation and outcome evaluation during slot machine play', *Neuroimage*, 232, 117874. doi: 10.1016/j.neuroimage.2021.117874

Fuchs, T. (2010) 'Subjectivity and intersubjectivity in psychiatric diagnosis', *Psychopathology*, 43(4), pp. 268–274. doi: 10.1159/000315126

Fumerton, R.A. (1980) 'Induction and reasoning to the best explanation', *Philosophy of Science*, 47, pp. 589–600. doi: 10.1086/288959

Fusar-Poli, P., Estradé, A., Stanghellini, G., Venables, J., Onwumere, J., Messas, G., Gilardi, L., Nelson, B., Patel, V., Bonoldi, I., Aragona, M., Cabrera, A., Rico, J., Hoque, A., Otaiku, J., Hunter, N., Tamelini, M. G., Maschião, L. F., Puchivailo, M. C., Piedade, V. L., Kéri, P., Kpodo, L., Sunkel, C., Bao, J., Shiers, D., Kuipers, E., Arango, C., and Maj, M. (2022) 'The lived experience of psychosis: A bottom-up review co-written by experts by experience and academics' *World Psychiatry*, 21 (2), pp. 168–188. doi: 10.1002/wps.20959

Gabbay, D. and Woods, J. (2005) *The Reach of Abduction: Insight and Trial. Vol. 2.* Amsterdam: North-Holland.

Gallagher, S. (2009) 'Two problems of intersubjectivity', *Journal of Consciousness Studies*, 16 (6–8), pp. 289–308.

García-Gutiérrez, M.S., Navarrete, F., Sala, F., Gasparyan, A., Austrich-Olivares, A. and Manzanares, J. (2020) 'Biomarkers in psychiatry: concept, definition, types and relevance to the clinical reality', *Frontiers in psychiatry*, 11, 432. doi: 10.3389/fpsyt.2020.00432

Garb, H. N. (1996). 'The representativeness and past-behavior heuristics in clinical judgment', *Professional Psychology: Research and Practice*, 27 (3), pp. 272–277. doi:10.1037/0735-7028.27.3.272

Garden, G. (2005) 'Physical examination in psychiatric practice', *Advances in Psychiatric Treatment*, 11 (2), pp. 142–149. doi: 10.1192/apt.11.2.142

Garfinkel, A. (1981) *Forms of explanation*. New Haven: Yale University Press.

Gelder, M.G., Lopez-Ibor, J.J. and Andreasen, N.C. (ed.) (2000) *The New Oxford Textbook of Psychiatry.* New York: Oxford University Press

Geugies, H., Groenewold, N.A., Meurs, M., Doornbos, B., de Klerk-Sluis, J.M., van Eijndhoven, P., Roest, A.E. and Ruhé, H. G. (2022). 'Decreased reward circuit connectivity during reward anticipation in major depression', *NeuroImage: Clinical*, 36, 103226, 10.1016/j.nicl.2022.103226.

Ghaemi, S.N. (2007) 'Feeling and time: the phenomenology of mood disorders, depressive realism, and existential psychotherapy', *Schizophrenia Bulletin*, 33 (1), pp. 122–130. doi: /10.1093/schbul/sbl061

Ghaemi, S.N. (2018) 'After the failure of DSM: Clinical research on psychiatric diagnosis', *World Psychiatry*, 17 (3), pp. 301–302. doi: 10.1002/wps.20563

Giere, R.N. (1988) *Explaining Science: A Cognitive Approach*. University of Chicago Press.

Giere, R.N. (1999). 'Using models to represent reality', in Cañas, A.G. (ed.) *Model-based reasoning in scientific discovery*. Boston, MA: Springer, pp.41-57.

Giesen, D. (1988) *International Medical Malpractice Law: A Comparative Law Study of Civil Liability Arising from Medical Care.* Leiden: Brill.

Glannon, W. (2022) 'Biomarkers in Psychiatric Disorders', *Cambridge Quarterly of Healthcare Ethics*, 31 (4), pp. 444–452. doi: 10.1017/S0963180122000056

Godfrey-Smith, P. (2003) *Theory and Reality: An Introduction to the Philosophy of Science.* Chicago: University of Chicago Press.

Godfrey-Smith, P. (2006) 'The strategy of model-based science', *Biology and Philosophy*, 21 (5), pp. 725–740. doi: 10.1007/s10539-006-9054-6

Godfrey-Smith, P. (2009) 'Models and fictions in science. Philosophical Studies', 143 (1), pp. 101–116. doi: 10.1007/s11098-008-9313-2

Goldman, A.I. (1986) *Epistemology and Cognition*. Harvard University Press.

Goldman, A.I. (1999). *Knowledge in a Social World*. Oxford University Press.

Goldman, A.I. (2000) 'Can science know when you're conscious?' *Journal of Consciousness Studies*, 7 (5), pp. 3–22.

Goldman, A.I. (2006) *Simulating minds: The philosophy, psychology, and neuroscience of mindreading*. Oxford: Oxford University Press.

Goldman, R.N. and Greenberg, L.S. (2015) *Case Formulation in Emotion-Focused Therapy: Co-Creating Clinical Maps for Change.* Washington, DC: American Psychological Association.

Goodwin, R.M. (1967) 'A Growth Cycle', in Feinstein, C.H. (ed.), *Socialism, Capitalism and Economic Growth*. Cambridge: Cambridge University Press, 165–170.

Graham, G. (2013) *The Disordered Mind: An Introduction to Philosophy of Mind and Mental Illness.* Routledge.

Griesinger, W. (1861) *Die Pathologie und Therapie der psychischen Krankheiten.* Stuttgart: Adolph Krabbe.

Groves, M., O'Rourke, P. and Alexander, H. (2003) 'The clinical reasoning characteristics of diagnostic experts', *Medical Teacher*, 25 (3), pp. 308–313. doi: 10.1080/0142159031000100427

Gupta, M., Potter, N. and Goyer, S. (2019) 'Diagnostic reasoning in psychiatry: acknowledging an explicit role for intersubjective knowing', *Philosophy, Psychiatry, & Psychology*, 26 (1), pp. 49–64. doi: 10.1353/ppp.2019.0003

Hamscher, L., Console, J. and de Kleer, J. (ed.) (1992) *Readings in Model-Based Diagnosis.* San Mateo: Morgan Kaufmann.

Happell, B., O'Donovan, A., Sharrock, J., Warner, T. and Gordon, S. (2022) 'Understanding the impact of expert by experience roles in mental health education', *Nurse Education Today*, 111, pp. 105324. doi: 10.1016/j.nedt.2022.105324

Harjai, P.K. and Tiwari, R. (2009) 'Model of critical diagnostic reasoning: Achieving expert clinician performance', *Nursing Education Perspectives*, 30 (5), pp. 305–311.

Harper, D. (n.d.). Etymology of apophatic. Online Etymology Dictionary. Retrieved 25 March 2023, from https://www.etymonline.com/word/apophatic [Accessed 25 March 2023].

Harré, R. and Madden, E.H. (1975) *Causal Powers: A Theory of Natural Necessity.* Oxford: Basil Blackwell.

Harvey, P.O., Pruessner, J., Czechowska, Y. and Lepage, M. (2007) 'Individual differences in trait anhedonia: a structural and functional magnetic resonance imaging study in non-clinical subjects', *Molecular Psychiatry*, 12, pp. 767–775. doi: 10.1038/sj.mp.4002021

Hawton, K., Saunders, K., Topiwala, A. and Haw, C. (2013) 'Psychiatric disorders in patients presenting to hospital following self-harm: a systematic review', *Journal of Affective Disorders*, 151(3), pp 821–830. doi: 10.1016/j.jad.2013.08.020

He, R., Yan, X., Guo, J., Xu, Q., Tang, B. and Sun, Q. (2018) 'Recent advances in biomarkers for Parkinson's disease', *Frontiers in aging neuroscience*, 10, pp. 305. doi: 10.3389/fnagi.2018.00305

Heal, J. (2003) *Mind, reason, and imagination: Selected essays in philosophy of mind and language*. Cambridge: Cambridge University Press.

Heinz, A. (2017) *A New Understanding of Mental Disorders: Computational Models for Dimensional Psychiatry.* Cambridge: MIT Press.

Helzer, J.E., Kraemer, H.C., Krueger, R.F., Wittchen, H.U., Sirovatka, P.J. and Regier, D.A. (ed.) (2009) *Dimensional Approaches in Diagnostic Classification: Refining the Research Agenda for DSM-V.* Washington, DC: American Psychiatric Pub.

Hengartner, M.P., Lehmann, S.N. (2017) 'Why Psychiatric Research Must Abandon Traditional Diagnostic Classification and Adopt a Fully Dimensional Scope: Two Solutions to a Persistent Problem' *Front Psychiatry*, 8, 101. doi: 10.3389/fpsyt.2017.00101

Hernando, A., Villuendas, D., Vesperinas, C., Abad, M., Plastino, A. (2010) 'Unravelling the size distribution of social groups with information theory in complex networks', *The European Physical Journal B*, 76, pp. 87–97. doi: 10.1140/epjb/e2010-00216-1

Hesslow, G. (1983) 'Explaining differences and weighting causes', *Theoria*, 49, pp. 87–111. doi: 10.1111/j.1755-2567.1983.tb00209.x

Hill, A.B. (1965) 'The environment and disease: Association or causation?', *Proceedings of the Royal Society of Medicine*, 58 (5), pp. 295–300.

Hirsch, C.R. and Mathews, A. (2012) 'A cognitive model of pathological worry', *Behaviour Research and Therapy*, 50 (10), pp. 636–646. doi: 10.1016/j.brat.2012.06.007

Hirsch, C.R., Mathews, A., Lequertier, B., Perman, G. and Hayes, S. (2013) 'Characteristics of worry in generalized anxiety disorder', *Journal of Behavior Therapy and Experimental Psychiatry*, 44 (4), pp. 388–395. doi: 10.1016%2Fj.jbtep.2013.03.004

Hodges, J.R. (2017) *Cognitive Assessment for Clinicians.* Oxford: Oxford University Press.

Horn, L.R. and Wansing, H. (2020) 'Negation', in Zalta, E.N. (ed.) The Stanford Encyclopedia of Philosophy. Available at: <https://plato.stanford.edu/archives/spr2020/entries/negation/> (Accessed: 22.09.23).

Hohwy, J. (2013) *The predictive mind.* Oxford: Oxford University Press.

Hoyer, J., Becker, E.S., and Roth, W.T. (2001) 'Characteristics of worry in GAD patients, social phobics, and controls', *Depression and anxiety*, 13 (2), pp. 89–96. doi: 10.1002/da.1021

Humphries, S.R. (1988) 'Munchausen Syndrome: Motives and the Relation to Deliberate Self-Harm', *British Journal of Psychiatry*, 152 (3), pp. 416–417. doi: 10.1192/bjp.152.3.416

Hurley, S. (2008) 'Understanding Simulation', *Philosophy and Phenomenological Research*, 77 (3), pp. 755–774. doi: 0.1111/j.1933-1592.2008.00220.x

Huys, Q.J., Pizzagalli, D.A., Bogdan, R., and Dayan, P. (2013), 'Mapping anhedonia onto reinforcement learning: a behavioural meta-analysis', *Biology of mood and anxiety disorders*, 3 (1), pp. 1–16. doi: 10.1186/2045-5380-3-12

Huys, Q.J., Maia, T.V., and Frank, M.J. (2016) 'Computational psychiatry as a bridge from neuroscience to clinical applications', *Nature Neuroscience*, 19 (3), pp. 404–413. doi: 10.1038/nn.4238

Insel, T. R. and Quirion, R. (2005) 'Psychiatry as a clinical neuroscience discipline', *Journal of the American Medical Association*, 294, pp. 2221–2224. doi: 10.1001/jama.294.17.2221

Insel, T.R. and Cuthbert, B.N. (2015) 'Brain disorders? Precisely', *Science*, 348, pp. 499–500. doi: 10.1126/science.aab2358

Jaspers K. [1913] (1963) *General Psychopathology. Translation of the German 7th edition.* J. Hoenig, J., and Hamilton, M.W. (trans). Manchester: Manchester University Press.

Jefferson, A. (2020) What does it take to be a brain disorder? *Synthese*, 197 (1), pp. 249–262. doi: 10.1007/s11229-018-1784-x

Johnson, C.P. and Myers, S.M. (2007) 'Identification and evaluation of children with autism spectrum disorders', *Pediatrics*, 120 (5), pp. 1183–1215. doi: 10.1542/peds.2007-2361

Johnson, D.K. (2019). 'Inference to the Best Explanation and Avoiding Diagnostic Error' in Allhoff, F., Sandra Borden, S. (ed.) *Ethics and Error in Medicine*, pp. 243–261. London: Routledge.

Johnstone, L. and Dallos, R. (2013) *Formulation in Psychology and Psychotherapy: Making Sense of People's Problems.* London: Routledge.

Johnson-Laird, P.N., (2010) 'Mental models and human reasoning', *Proceedings of the National Academy of Sciences*, 107 (43), pp. 18243–18250. doi: 10.1073/pnas.1012933107

Jones, S., Howard, L., and Thornicroft, G. (2008) 'Diagnostic overshadowing: worse physical health care for people with mental illness', *Acta Psychiatrica Scandinavia*, 118, pp. 169–171. doi: 10.1111/j.1600-0447.2008.01211.x

Kapur, S., Phillips, A.G. and Insel, T.R. (2012) 'Why has it taken so long for biological psychiatry to develop clinical tests and what to do about it?', *Molecular Psychiatry*, 17 (12), pp.1174-1179. doi: 10.1038/mp.2012.105

Kayser, M.S., Titulaer, M.J., Gresa-Arribas, N. and Dalmau, J. (2013) 'Frequency and Characteristics of Isolated Psychiatric Episodes in Anti–N-Methyl-d-Aspartate Receptor Encephalitis', *JAMA Neurology*, 70 (9), pp.1133-1139. doi: 10.1001/jamaneurol.2013.3216

Kelley, T.L. (1927) *Interpretation of Educational Measurements*. Chicago: World Book Co.

Kendler, K.S. and Parnas, J. (ed.) (2012) *Philosophical Issues in Psychiatry II: Nosology*. Oxford: Oxford University Press.

Kennerley, H., Kirk, J. and Westbrook, D. (2016) 'Assessment and Formulation', in Kennerley, H., Kirk, J. and Westbrook, D. (ed.) *An Introduction to Cognitive Behaviour Therapy: Skills and Applications*, New York: Sage, pp. 37–62.

Kind, A. (2023) 'How does the psychiatrist know?' *Philosophy and the Mind Sciences*, 4. doi: 10.33735/phimisci.2023.9391

Klein, D.F. (1974) 'Endogenomorphic depression: A conceptual and terminological revision', *Archives of General Psychiatry*, 31, pp. 447–454. doi: 10.1001/archpsyc.1974.01760160005001

Klein, D. (1984) 'Depression and Anhedonia', in Clark, C. and Fawcett, J. (ed.) *Anhedonia and Affect Deficit States*. New York, PMA Publishing, pp. 1–14.

Klonsky, E.D., Victor, S.E. and Saffer, B.Y. (2014) 'Nonsuicidal self-injury: what we know, and what we need to know', *Canadian Journal of Psychiatry*, 59 (11), pp. 565–568. doi: 10.1177/070674371405901101

Knuuttila, T. and Loettgers, A. (2016) 'Model templates within and between disciplines: from magnets to gases–and socio-economic systems', *European Journal for Philosophy of Science*, 6 (3), pp. 377–400. doi: 10.1007/s13194-016-0145-1

Koehler, D.J., Brenner, L. and Griffin, D. (2002) 'The calibration of expert judgment: Heuristics and biases beyond the laboratory', in Gilovich, T., Griffin, D. and Kahneman, D. (ed.) *Heuristics and Biases: The Psychology of Intuitive Judgment*. Cambridge: Cambridge University Press, pp. 686–715.

Kotov, R., Krueger, R. F., Watson, D., Achenbach, T. M., Althoff, R. R., Bagby, R. M., Brown, T.A., Carpenter, W.T, , Caspi, A. , Clark, L.E., Eaton, N.R. Forbes, M.K., Forbush, K.T., Goldberg, D., Hasin D., Hyman, S.E., Ivanova, M.Y., Lynam, D.R., Markon, K., Miller, J.D., Moffitt, T.E., Morey, L.C., Mullins-Sweatt, S.N., Ormel, J., Patrick C.J., Regier, D.A., Rescorla, L., Ruggero, C.J., Samuel, D.B., Sellbom,

M., Simms, L.J., Skodol, A.E., Slade, T., South, S.C., Tackett J.L., Waldman, I.D., Waszczuk, M.A., Widiger, T.A., Wright, A.G.C., Zimmerman, M. (2017). 'The Hierarchical Taxonomy of Psychopathology (HiTOP): A dimensional alternative to traditional nosologies', *Journal of Abnormal Psychology*, 126 (4), pp. 454–477. doi: 10.1037/abn0000258

Kraepelin E. (1919) *Dementia praecox and paraphrenia*. Barclay R.M., (trans.). Chicago: Chicago Medical Books.

Kringelbach, M.L. (2005), 'The human orbitofrontal cortex: linking reward to hedonic experience', *Nature Reviews Neuroscience*, 6, pp. 691–702. doi: 10.1038/nrn1747

Krueger, R.F. and Bezdjian, S. (2009) 'Enhancing research and treatment of mental disorders with dimensional concepts: toward DSM-V and ICD-11', *World Psychiatry: Official Journal of the World Psychiatric Association*, 8 (1), pp. 3–6. doi: 10.1002/j.2051-5545.2009.tb00197.x

Kuhlmann, S.R., Walter, H. and Schläpfer, T.E. (2013) 'Neurobiologie der Anhedonie: Zur Pathophysiologie eines bedeutenden Symptoms der depressiven Störung', *Nervenarzt*, 84 (5), pp. 590–595. doi: 10.1007/s00115-012-3654-y

Kuruvilla, K. and Kuruvilla, A. (2010) 'Diagnostic formulation', *Indian Journal of Psychiatry*, 52 (1), pp. 78–82. doi: 10.4103/0019-5545.58905

Lambert, C., Da Silva, S., Ceniti, A.K., Rizvi, S.J., Foussias, G. and Kennedy, S.H. (2018) 'Anhedonia in depression and schizophrenia: A transdiagnostic challenge', *CNS Neuroscience & Therapeutics*, 24 (7), pp. 615–623. doi: 10.1111/cns.12854

Langlois, F., Freeston, M.H. and Ladouceur, R. (2000a) 'Differences and similarities between obsessive intrusive thoughts and worry in a non-clinical population: Study 1', *Behaviour Research and Therapy*, 38 (2), pp.157-173. doi: 10.1016/s0005-7967(99)00028-5

Langlois, F., Freeston, M.H. and Ladouceur, R. (2000b) 'Differences and similarities between obsessive intrusive thoughts and worry in a non-clinical population: Study 2', *Behaviour Research and Therapy*, 38(2), pp.175-189. doi: 10.1177/1073191109344694

Lee, E. N. (1972) 'Plato on Negation and Not-Being in the Sophist', *The Philosophical Review*, 81(3), pp. 267–304. doi: 10.2307/2184327

Lee, H.-J., Lee, S.-H., Kim, H.-S., Kwon, S.-M., and Telch, M. J. (2003) 'A comparison of autogenous/reactive obsessions and worry in a nonclinical population.' *Poster presented at the annual meeting of the Association for the Advancement of Behavior Therapy*, Boston.

Leibenluft, E. and Stoddard, J. (2013) 'The developmental psychopathology of irritability', *Development and Psychopathology*, 25 (4), pp. 1473–1487. doi: 10.1017/S0954579413000722

Leplin, J. (2009). *A Theory of Epistemic Justification*. Berlin: Springer.

Lewis, D. (2001) 'Truthmaking and Difference-Making'. *Noûs*, 35 (4), pp. 602–615. doi: 10.1111/0029-4624.00354

Liang, S., Wu, Y., Hanxiaoran, L., Greenshaw, A. J., & Li, T. (2022) 'Anhedonia in Depression and Schizophrenia: Brain Reward and Aversion Circuits', *Neuropsychiatric Disease and Treatment*, 18, pp. 1385–1396. doi: 10.2147/NDT.S367839

Lindberg, J., Marco, B. and Klas, S. (2013) 'A case of celiac disease with hallucinations', *Neurology Clinical Practice*, 3 (5), pp. 446–447. doi: 10.1212/CPJ.0b013e3182a78f37

Lipton, P. (1991). *Inference to the Best Explanation (2nd ed.)*. London: Routledge.

Lipton, P. (2017). Inference to the Best Explanation, in Newton-Smith, W.H. (ed.) *A Companion to the Philosophy of Science*. (pp. 184–193). Chichester: Wiley.

Lotka, A. J. (1910) 'Contribution to the Theory of Periodic Reaction', *Journal of Physical Chemistry*, 14 (3), pp. 271–274.

Lotka, A. J. (1925). *Elements of Physical Biology.* Williams and Wilkins.

Loveday, T., Wiggins, M., Festa, M., Schell, D. and Twigg, D. (2013) 'Pattern Recognition as an Indicator of Diagnostic Expertise', in Latorre Carmona, p., Sánchez, J. S., and Fred, A.L.N. (ed.) *Pattern Recognition – Applications and Methods.* Berlin: Springer, pp. 1–11.

Console, L. and Torasso, P. (1991) 'On the co-operation between abductive and temporal reasoning in medical diagnosis', *Artificial Intelligence in Medicine*, 3 (6), pp. 291–311. doi: 10.1016/0933-3657(91)90002-S

Lutz, K. and Widmer, M. (2014) 'What can the monetary incentive delay task tell us about the neural processing of reward and punishment? ', *Neuroscience and Neuroeconomics*, 3, pp. 33–45. doi: 10.2147/NAN.S38864

Friedman, M.O., Pham, S., Abou Abdallah, A., Guelfi, T. and Rouillon, F. (2008) 'Personnalité borderline, automutilations et suicide : revue de la littérature [Borderline personality disorder, self-mutilation and suicide: literature review]', *Encephale*, 34 (5), pp. 452–458. doi: 10.1016/j.encep.2007.10.007

Mackie, J.L. (1974) *The Cement of the Universe*. Oxford: Oxford University Press.

Maclin, J.M.A., Wang, T., Xiao, S. (2019) 'Biomarkers for the diagnosis of Alzheimer's disease, dementia Lewy body, frontotemporal dementia and vascular dementia', *General Psychiatry*, 32 (1), e100054. doi: 10.1136/gpsych-2019-100054

Madhusoodanan, S., Danan, D., Brenner, R. and Bogunovic, O. (2004) 'Brain tumor and psychiatric manifestations: a case report and brief review', *Annals of Clinical Psychiatry*, 16 (2), pp. 111–113. doi: 10.1080/10401230490453770

Mamede, S., Schmidt, H. G. and Rikers, R. (2007) 'Diagnostic errors and reflective practice in medicine', *Journal of Evaluation in Clinical Practice*, 13 (1), pp. 138–145. doi: 10.1111/j.1365-2753.2006.00638.x

Marcum, J.A. (2012) 'An integrated model of clinical reasoning: Dual-process theory of cognition and metacognition', *Journal of Evaluation in Clinical Practice*, 18(5), pp. 954–961. doi: 10.1111/j.1365-2753.2012.01900.x

Martins-de-Souza, D., (2013) 'Biomarkers for psychiatric disorders: where are we standing? ' *Disease markers*, 35 (1), pp.1-2. doi: 10.1155/2013/321071

Marquand, A.F., Kia, S.M., Zabihi, Wolfers, T., Buitelaar, J.K., Beckmann, C.F. (2019) 'Conceptualizing mental disorders as deviations from normative functioning', *Molecular Psychiatry*, 24, pp. 1415–1424. doi: 10.1038/s41380-019-0441-1

Mason, O.J, and Brady, F. (2009) 'The psychotomimetic effects of short-term sensory deprivation', *Journal of Nervous and Mental Disease*, 197 (10), pp.783-785. doi: 10.1097/NMD.0b013e3181b9760b

Matthewson, J. and Weisberg, M. (2009) 'The structure of tradeoffs in model building', *Synthese*, 170 (1), pp. 169–190. doi: 10.1007/s11229-008-9366-y

McHugh, P.R. and Slavney, P.R. (1998) *The Perspectives of Psychiatry*. Baltimore: Johns Hopkins University Press.

McMullin, E. (1985) 'Galilean Idealization', *Studies in History and Philosophy of Science*, 16, pp. 247–273. doi: 10.1016/0039-3681(85)90003-2

McWilliams, N. (1999). *Psychoanalytic case formulation*. New York: Guilford Press.

Mebius, A., Kennedy, A. G. and Howick, J. (2016) 'Research gaps in the philosophy of evidence-based medicine', *Philosophy Compass*, 11 (11), pp. 757–771. doi: doi.org/10.1111/phc3.12352

Menzies, P. and Beebee, H. (2020) 'Counterfactual Theories of Causation', in Zalta, E.N. (ed.) The Stanford Encyclopedia of Philosophy. Available at: https://plato.stanford.edu/archives/win2020/entries/causation-counterfactual/ (Accessed: 22. 09. 2023).

Meyer, V. and Turkat, I.D. (1979) 'Behavioral analysis of clinical cases', *Journal of Behavioral Assessment*, 1, pp. 259–270. doi: 10.1007/BF01321368

Michel, M. and Matthias, M. (2019) 'The Mismeasure of Consciousness: A Problem of Coordination for the Perceptual Awareness Scale', *Philosophy of Science*, 86 (5), pp. 1239–1249. doi: 10.1086/705509

Miller, D.B. and O'Callaghan, J.P. (2015) 'Biomarkers of Parkinson's disease: present and future', *Metabolism*, 64 (3), pp. 40–46. doi: https://doi.org/10.1016/j.metabol.2014.10.030

Miller, G.A. (2010) 'Mistreating psychology in the decades of the brain', *Perspectives on Psychological Science*, 5 (6), pp. 716–743. doi: 10.1177/1745691610388774

Mittal, M. and Khan, S. (2010) 'Starvation causes acute psychosis due to anterior thalamic infarction', *Southern Medical Journal*, 103 (7), pp. 701–703. doi: 10.1097/SMJ.0b013e3181e1e2f3

Montague, P.R., Dolan, R.J., Friston, K.J. and Dayan, P. (2012) 'Computational psychiatry', *Trends in Cognitive Sciences*, 16 (1), pp. 72–80. doi: 10.1016/j.tics.2011.11.018

Monteiro, S. M., and Norman, G. (2013) 'Diagnostic reasoning: where we've been, where we're going', *Teaching and learning in medicine*, 25 (1), pp. 26–32. doi: 10.1080/10401334.2013.842911

Morrison, M. and Morgan, M. S. (1999) 'Models as mediating instruments' in Morgan, M.S. and Morrison, M. (ed.) *Models as Mediators: Perspectives on Natural and Social Science.* Cambridge: Cambridge University Press, pp. 10–37.

Morrison, M. (1999) 'Models as autonomous agents' in Morgan, M.S. and Morrison, M. (ed.) *Models as Mediators: Perspectives on Natural and Social Science.* Cambridge: Cambridge University Press, pp. 38–65.

Morrison, A.P. (2005). 'Psychosis and the phenomenon of unwanted intrusive thoughts' in Clark, D.A. (ed.), *Intrusive thoughts in clinical disorders: Theory, research, and treatment*. New York: Guilford, pp. 175–198.

Mu, E. and Kulkarni, J. (2022) 'Hormonal contraception and mood disorders', *Australian Prescriber*, 45 (3), pp. 75–79. doi: 10.18773/austprescr.2022.025

Mueller, A.E. and Segal, D.L. (2014) 'Structured versus semistructured versus unstructured interviews', in Cautin, R.L., Lilenfeld, S.O. (ed.) *The Encyclopaedia of Clinical Psychology*. New York: John Wiley & Sons.

Murphy, D. (2012). Psychiatry in the scientific image. Cambridge, MA: MIT Press.

Murphy, D. (2020) 'Philosophy of Psychiatry', Zalta, E.N. (ed.), *The Stanford Encyclopedia of Philosophy.* Retrieved from https://plato.stanford.edu/archives/fall2020/entries/psychiatry/ (Accessed 22.09.2023).

Murphy, M.J. and Peterson, M.J. (2015) 'Sleep Disturbances in Depression', *Sleep Medicine Clinics*, 10 (1), pp.17-23. doi: 10.1016/j.jsmc.2014.11.009

Myers, R.H. (2004). 'Huntington's disease genetics', *NeuroRx: The Journal of the American Society for Experimental NeuroTherapeutics*, 1(2), pp. 255–262. doi: 10.1602/neurorx.1.2.255

Myerson, A. (1920). *The Nervous Housewife.* Boston: Brown & Little.

Nakash, O., Alegría, M. (2013) 'Examination of the role of implicit clinical judgments during the mental health intake', *Qualitative Health Research*, 23 (5), pp. 645–654. doi: 10.1177/1049732312471732

National Academies of Sciences, Engineering, and Medicine (NASEM) (2019) 'Diagnosis and assessment of traumatic brain injury' in NASEM *Evaluation of the disability determination process for traumatic brain injury in veterans.* Washington D.C.: National Academies Press, pp. 23–56.

National Institute of Mental Health (2013) *Research Domain Criteria (RDoC)*. Available at: <https://www.nimh.nih.gov/research-priorities/rdoc/index.shtml> [Accessed 01. 09. 2021].

National Institute of Mental Health (2018) *RDoC changes to the matrix (CMAT) workgroup update: proposed positive valence domain revisions*. Available at: https://www.nimh.nih.gov/about/advisory-boards-and-groups/namhc/reports/cmat-pvs-report-508_157003.pdf. [Accessed 01. 09. 2021].

Nordgaard, J., Revsbech, R., Sæbye, D., and Parnas, J. (2012). 'Assessing the diagnostic validity of a structured psychiatric interview in a first-admission hospital sample', *World Psychiatry*, 11 (3), pp. 181–185. doi: 10.1002/j.2051-5545.2012.tb00128.x

Odenbaugh, J. (2003) 'Complex systems, trade-offs, and theoretical population biology: Richard Levin's "strategy of model building in population biology" revisited', *Philosophy of Science*, 70 (5), pp.1496-1507. doi: 10.1086/377425

Ohayon, M.M., Priest, R.G., Caulet, M. and Guilleminault, C. (1996) Hypnagogic and hypnopompic hallucinations: pathological phenomena? *British Journal of Psychiatry*, 169 (4), pp. 459–467. doi: 10.1192/bjp.169.4.459.

Ohayon, M.M. (2002) 'Epidemiology of insomnia: What we know and what we still need to learn', *Sleep Medicine Reviews*, 6, pp. 97–111. doi: 10.1053/smrv.2002.0186

Ohayon, M.M. (2005) 'Relationship between chronic painful physical condition and insomnia', *Journal of Psychiatric Research*, 39, pp.151-159. doi: 10.1016/j.jpsychires.2004.07.001

Olbert, C.M. and Gala, G.J. (2015) 'Supervenience and psychiatry: Are mental disorders brain disorders?', *Journal of Theoretical and Philosophical Psychology*, 35 (4), pp. 203–219. Doi: 10.1037/te00000023

Orzack, S.H. (2005) 'Discussion: What, if Anything, is "the Strategy of Model Building in Population Biology?" A Comment on Levins (1966) and Odenbaugh (2003)', *Philosophy of Science*, 72 (3), pp.479-485. doi: 10.1086/498475

Orzack, S.H. (2012) 'The philosophy of modelling or does the philosophy of biology have any use?' *Philosophical Transactions of the Royal Society B: Biological Sciences*, 367, pp. 170–180. doi: 10.1098/rstb.2011.0265

Papineau, D. (1994) 'Mental disorder, illness and biological disfunction', *Royal Institute of Philosophy Supplements*, 37, pp. 73–82. doi: 10.1017/S135824610000998X

Park, S.C., Kim, J.M., Jun, T.Y., Lee, M.S., Kim, J.B., Yim, H.W., and Park, Y.C. (2017) 'How many different symptom combinations fulfil the diagnostic criteria for major depressive disorder? Results from the CRESCEND study', *Nordic Journal of Psychiatry*, 71 (3), 217–222. doi: 10.1080/08039488.2016.1265584

Parnas, J., Sass, L.A. and Zahavi, D. (2013) 'Rediscovering psychopathology: The epistemology and phenomenology of the psychiatric object', *Schizophrenia Bulletin*, 39 (2), pp. 270–277. doi: 10.1093/schbul/sbs153

Pauen, M. (2012) 'The second-person perspective', *Inquiry*, 55 (1), pp. 33–49. doi: 10.1080/0020174x.2012.643623

Perälä, M. (2020) 'Affirmation and Denial in Aristotle's De interpretatione', *Topoi*, 39, pp. 645–656. doi: 10.1007/s11245-019-09669-y

Peirce, C.S. [1878] (1932) 'Deduction, induction, and hypothesis', in Hartshorne, C and Weiss, P. (ed.), *Collected Papers of Charles Sanders Peirce, vol. 2*. Cambridge: Harvard University Press, pp. 619–644.

Peirce, C.S. [1903] (1932) 'Harvard Lectures on Pragmatism' in Hartshorne, C. and Weiss, P. (ed.), *Collected Papers of Charles Sanders Peirce, vol. 5*. Cambridge: Harvard University Press, 13–131.

Pelizza, L. and Ferrari, A. (2009) 'Anhedonia in schizophrenia and major depression: state or trait? ', *Annals of General Psychiatry*, 8, pp. 22–30. doi: 10.1186/1744-859X-8-22

Pi, Y., Liao, W., Liu, M. and Lu, J. (2008) 'Theory of Cognitive Pattern Recognition', in Yin, P.-Y. (ed.) *Pattern Recognition Techniques, Technology and Applications*. London: InTech, pp. 433–462.

Place, U.T. (1956) 'Is Consciousness a Brain Process?', *British Journal of Psychology*, 47, pp. 44–50. doi: /10.1111/j.2044-8295.1956.tb00560.x

Plato (1937). *The Sophist*. Translated by Jowett, B. New York: Random House.

Pollock, J. L. (1986) *Contemporary theories of knowledge*. Totowa, NJ: Rowman and Littlefield.

Poole, R. and Higgo, R. (2017) *Psychiatric Interviewing and Assessment, 2nd ed.* Cambridge: Cambridge University Press.

Popper, K.R. [1935] (1959) The Logic of Scientific Discovery. London: Hutchinson & Sons.

Power, B.D., Nguyen, T., Hayhow, B. and Looi, J.C.L (2016) 'Neuroimaging in psychiatry: an update on neuroimaging in the clinical setting', *Australasian Psychiatry*, 24 (2), pp. 157–163. doi: 10.1177/1039856215618525

Joyce, P.R., Light, K.J., Rowe, S.L., Cloninger, C.R., Kennedy, M.A. (2010) 'Self-mutilation and suicide attempts: relationships to bipolar disorder, borderline personality disorder, temperament and character', *Australian and New Zealand Journal of Psychiatry*, 44 (3), 250–257. doi: 10.3109/00048670903487159

Price, B.H., Adams, R.D. and Coyle, J.T. (2000) 'Neurology and psychiatry: closing the great divide', *Neurology*, 54 (1), p. 8–14. doi: 10.1212/wnl.54.1.8

Rachman, S. (1978) 'An anatomy of obsessions', *Behavioural Analysis and Modification*, 2, pp. 235–278.

Rachman, S. (1981) 'Unwanted intrusive cognitions', *Advances in Behaviour Research and Therapy*, 3, pp. 89–99. doi: 10.1016/0146-6402(81)90007-2

Rachman, S.J. (1997) 'A cognitive theory of obsessions', *Behaviour Research and Therapy*, 35, pp. 793–802. doi: 10.1016/s0005-7967(97)00040-5

Rachman, S.J. (1998) 'A cognitive theory of obsessions: Elaborations', *Behaviour Research and Therapy*, 36, pp. 385–401. doi: 10.1016/s0005-7967(97)10041-9

Rachman, S. (2003) *The Treatment of Obsessions*. Oxford: Oxford University Press.

Rapp, P.E., Cellucci, C., Darmon, D. and Keyser, D. (2022) 'Cautionary Observations Concerning the Introduction of Psychophysiological Biomarkers into Neuropsychiatric Practice', *Psychiatry International*, 3 (2), pp. 181–205. doi: 10.3390/psychiatryint3020015

Ratcliffe, M. (2015) *Experiences of depression: A study in phenomenology*. Oxford: Oxford University Press.

Reed, G.M., First, M.B., Kogan, C.S., Hyman, S.E., Gureje, O., Gaebel, W., Maj, M., Stein, D.J., Maercker, A., Tyrer, P., Claudino, A., Garralda, E., Salvador-Carulla,

L., Ray, R., Saunders, J.B., Dua, T., Poznyak, V., Medina-Mora, M.E., Pike, K.M., Ayuso-Mateos, J.L., Kanba, S., Keeley, J.W., Khoury, B., Krasnov, V.N., Kulygina, M., Lovell, A.M., de Jesus Mari J., Maruta, T., Matsumoto, C., Rebello, T.J., Roberts, M.C., Robles, R., Sharan, P., Zhao, M., Jablensky, A., Udomratn, P., Rahimi-Movaghar, A., Rydelius, P.A., Bährer-Kohler, S., Watts, A.D., Saxena, S. (2019) 'Innovations and changes in the ICD-11 classification of mental, behavioural and neurodevelopmental disorders'. *World Psychiatry*, 18 (1), pp. 3–19. doi: 10.1002/wps.20611

Regier D.A., Kuhl E.A, Kupfer D.J. (2013) 'The DSM-5: Classification and criteria changes', *World Psychiatry*, 12 (2), pp. 92–98. doi: 10.1002/wps.20050

Reiter, R. (1987) A theory of diagnosis from first principles. *Artificial Intelligence*, 32 (1), 57–95. doi: 10.1016/0004-3702(87)90062-2

Reitz, C. and Mayeux, R. (2014) 'Alzheimer disease: epidemiology, diagnostic criteria, risk factors and biomarkers', *Biochemical Pharmacology*, 88 (4), pp. 640–651. doi: 10.1016/j.bcp.2013.12.024

Rescorla, R.A. (1988) 'Pavlovian conditioning: It's not what you think it is', *American Psychologist*, 43 (3), 151–160. doi: 10.1037//0003-066x.43.3.151.

Reum, J.C., McDonald, P.S., Ferriss, B.E., Farrell, D.M., Harvey, C.J. and Levin, P.S. (2015) 'Qualitative network models in support of ecosystem approaches to bivalve aquaculture', *ICES Journal of Marine Science*, 72 (8), pp. 2278–2288. doi: 10.1093/icesjms/fsv119

Reznek, L. (1998) 'On the epistemology of mental illness', *History and Philosophy of the Life Sciences*, 20 (2), pp. 215–232.

Ribot, T. (1897). *The Psychology of Emotions.* London: W. Scott.

Rizvi, S.J., Pizzagalli, D.A., Sproule, B.A. and Kennedy, S.H. (2016) 'Assessing anhedonia in depression: Potentials and pitfalls', *Neuroscience & Biobehavioural Reviews*, 65, pp.21-35. doi: 10.1016/j.neubiorev.2016.03.004

Rodríguez-López, Á., Rodríguez-Ortiz, E., Romero-Gonzalez, B. (2021) 'Non-suicidal self-injury in patients with eating disorders: nuclear aspects', *Colombia Médica*, 52 (1), e2044342, doi: 10.25100/cm.v51i4.4342

Rohrer J.D., Rosen H.J. (2013) 'Neuroimaging in frontotemporal dementia' *International Review of Psychiatry*, 25 (2), 221–229. doi: 10.3109/09540261.2013.778822

Romero-Sanchiz, P., Nogueira-Arjona, R., Godoy-Ávila, A., Gavino-Lázaro, A. and Freeston, M.H. (2017) 'Differences in clinical intrusive thoughts between obsessive–compulsive disorder, generalized anxiety disorder, and hypochondria', *Clinical Psychology and Psychotherapy*, 24(6), pp.1464-1473. doi: 10.1002/cpp.2107

Rosenhan, D.L. (1973) 'On being sane in insane places', *Science*, 179, pp.250-258. doi: 10.1002/cpp.2107

Roth, M., Huppert, F.A., Mountjoy, C.Q. and Tym, E. (1998) *CAMDEX-R: The Cambridge Examination for Mental Disorders of the Elderly – Revised.* Cambridge: Cambridge University Press.

Roughgarden, J., (1979) *Theory of Population Genetics and Evolutionary Ecology: An Introduction.* New York: Macmillan.

Royal Australian and New Zealand College of Psychiatrists (2014) 'Clinical Examinations: Formulation Guidelines for Trainees [Formulation Guidelines 0304]', retrieved from https://www.ranzcp.org/Files/Resources/Education/Training/Clinical_Examinations/Formulation_Guidelines_0304.pdf (Accessed 22. 09. 2023)

Royal College of Psychiatrists (2013) 'A Competency Based Curriculum for Specialist Core Training in Psychiatry: Core Training in Psychiatry CT1–CT3', retrieved from https://www.rcpsych.ac.uk/pdf/CT1-3_Curriculum_2013.pdf (Accessed 22. 09. 2023)

Royal College of Psychiatrists (2017) *Using Formulation in General Psychiatric Care: Good Practice.* London: Royal College of Psychiatrists.

Rubenstein, M. J. (2003) 'Unknow thyself: apophaticism, deconstruction, and theology after ontotheology', *Modern Theology*, 19 (3), pp. 387–417.

Ruscio, A.M. and Borkovec, T.D. (2004) 'Experience and appraisal of worry among high worriers with and without generalized anxiety disorder', *Behaviour Research and Therapy*, 42 (12), 1469–1482. doi: 10.1016/j.brat.2003.10.007

Ryan, P. (ed.) (2019) *Enhancing Clinical Case Formulation: Theoretical and Practical Approaches for Mental Health Practitioners*. Abingdon: Routledge.

Sachdev, P., Smith, J.S., and Cathcart, S. (2001) 'Schizophrenia-like psychosis following traumatic brain injury: a chart-based descriptive and case-control study', *Psychological Medicine*, 31 (2), 231–239. doi: 10.1017/s0033291701003336

Sadock, B.J. and Sadock, W.V. (2008) *Kaplan & Sadock's Concise Textbook of Clinical Psychiatry. 3rd Edition*, Philadelphia: Lippincott Williams & Wilkins

Sadock, B.J. and Sadock, V.A. (2014) *Kaplan and Sadock's Synopsis of Psychiatry: Behavioral Sciences/Clinical Psychiatry*. Philadelphia: Lippincott Williams & Wilkins.

Sargent, R. G. (2010) 'Verification and validation of simulation model', in Johnasson , B., Sanjay, J., Montoya-Torres, J., Yücesan, E., *Proceedings of the 2010 Winter Simulation Conference*. ACM Digital Library, 166–183

Schramme, T. (2013) 'On the autonomy of the concept of disease in psychiatry', *Frontiers in Psychology*, 4, 457. doi: 10.3389/fpsyg.2013.00457

Schultz, W. (2002) 'Getting formal with dopamine and reward', *Neuron*, 36 (2), pp. 241–263. doi: 10.1016/s0896-6273(02)00967-4

Schultz, W., Dayan, P. and Montague, P.R. (1997) 'A neural substrate of prediction and reward', *Science*, 275, pp. 1593–1599. doi: 10.1126/science.275.5306.1593

Schurz, G. (2008) 'Patterns of abduction', *Synthese*, 164, pp. 201–234. doi: 10.1007/s11229-007-9223-4

Schurz, G. (2009) 'Meliorative reliabilist epistemology: Where externalism and internalism meet', *Grazer Philosophische Studien*, 79, pp. 41–62. doi: 10.1163/18756735-90000856

Schurz, G. (2011) 'Truth-conduciveness as the primary epistemic justification of normative systems of reasoning', *Behavioral and Brain Sciences*, 34 (5), 266–267. doi: 10.1017/S0140525X11000537

Seriès, P. (ed.) (2020) *Computational Psychiatry: A Primer*. Cambridge: MIT Press.

Sescousse, G., Caldu, X., Segura, B., and Dreher, J. C. (2013) 'Processing of primary and secondary rewards: A quantitative meta-analysis and review of human functional neuroimaging studies', *Neuroscience and Biobehavioral Reviews*, 37, 681–696. doi: 10.1016/j.neubiorev.2013.02.002

Sherrington, C. (1906) *The Integrative Action of the Nervous System*. New York: C. Scribner's Sons.

Sibrava, N. and Borkovec, T.D. (2006) 'The cognitive avoidance theory of worry', in Davey, G.C.L. and Wells, A. (ed.), *Worry and its Psychological Disorders: Theory, Assessment and Treatment*. Hoboken: Wiley Publishing, pp. 217–237.

Silajdžić, E. and Björkqvist, M. (2018) 'A critical evaluation of wet biomarkers for Huntington's disease: current status and ways forward', *Journal of Huntington's Disease*, 7(2), pp. 109–135. doi: 10.3233/JHD-170273

Silva M.C.B.E., Brucki S.M.D. (2010) 'Multiple hallucinations due to brainstem injury: A case report' *Dementia & Neuropsychologia*, 4 (4), 353–355. doi: 10.1590/S1980-57642010DN40400016

Singhal, A., Ross, J., Seminog, O., Hawton, K. and Goldacre, M.J. (2014) 'Risk of self-harm and suicide in people with specific psychiatric and physical disorders: comparisons between disorders using English national record linkage', *Journal of the Royal Society of Medicine*, 107 (5), pp. 194–204. doi: 10.1590/S1980-57642010DN40400016

Skovlund C.W., Mørch L.S., Kessing L.V., Lidegaard Ø. (2016) 'Association of Hormonal Contraception With Depression', *JAMA Psychiatry*, 73 (11), pp.1154-1162. doi: 10.1590/S1980-57642010DN40400016

Smart, J.J.C. (1959) 'Sensations and Brain Processes'. *Philosophical Review*, 68, pp.141-156. doi: 10.2307/2182164

Sober, E. (1979) 'The Art and Science of Clinical Judgment', in Tristram, H., Stuart, E., Spicker, S.F., Towers, B. (ed.) *Clinical Judgment: A Critical Appraisal*. Berlin: Springer, pp. 29–44.

Spearman, C. (1904) 'General intelligence, objectively determined and measured', *American Journal of Psychology*, 15 (2), pp. 201–292. doi: 10.2307/1412107

Sperry, M.D. (1992) 'Demystifying the psychiatric case formulation', *Jefferson Journal of Psychiatry*, 10 (2), pp. 12–19. doi: 10.29046/JJP.010.2.002

Spitzer, R. (1975) 'On pseudoscience in science, logic in remission, and psychiatric diagnosis: A critique of Rosenhan's "On being sane in insane places"', *Journal of Abnormal Psychology*, 84 (5), pp. 442–452. doi: 10.1001/archpsyc.1976.01770040029007

Staddon, J.E. and Cerutti, D.T. (2003) 'Operant conditioning', *Annual Review of Psychology*, 54, pp. 115–144. doi: 10.1146/annurev.psych.54.101601.145124

Stanley, D.E. and Nyrup, R. (2020) 'Strategies in Abduction: Generating and Selecting Diagnostic Hypotheses', *The Journal of Medicine and Philosophy*, 45 (2), pp. 159–178. doi: 10.1093/jmp/jhz041

Stout, D.M., Simmons, A.N., Nievergelt, C.M., Minassian, A., Biswas, N., Maihofer, A. Risbrough, V.B., Baker, D.G. (2022) 'Deriving psychiatric symptom-based biomarkers from multivariate relationships between psychophysiological and biochemical measures' *Neuropsychopharmacology*, 47, 2252–2260. doi: 10.1038/s41386-022-01303-7

Stöber J. (1998) 'Worry, problem elaboration and suppression of imagery: the role of concreteness', *Behaviour research and therapy*, 36 (7–8), pp. 751–756. doi: 10.1016/s0005-7967(98)00027-8

Stöber, S., Tepperwien, S., Staak, M. (2000) 'Worrying Leads to Reduced Concreteness of Problem Elaborations: Evidence for the Avoidance Theory of Worry', *Anxiety, Stress, & Coping*, 13 (3), pp. 217–227. doi: 10.1080/10615800008549263

Stöber, J., and Borkovec, T.D. (2002) 'Reduced concreteness of worry in generalized anxiety disorder: Findings from a therapy study', *Cognitive Therapy and Research*, 26 (1), pp. 89–96. doi: 10.1023/A:1013845821848

Strevens, M. (2008) *Depth: An account of scientific explanation.* Cambridge: Harvard University Press.

Sullivan, E. and Khalifa, K. (2019) 'Idealizations and Understanding: Much Ado About Nothing?', *Australasian Journal of Philosophy*, 97 (4), pp. 673–689. doi: 10.1080/00048402.2018.1564337

Sunkel, C. and Sartor, C. (2022) 'Perspectives: involving persons with lived experience of mental health conditions in service delivery, development and leadership', *BJPsych Bulletin*, 46 (3), pp. 160–164. doi: 10.1080/00048402.2018.1564337

Sutton, R.S. and Barto, A.G. (2018) *Reinforcement Learning: An Introduction.* Cambridge, MA: MIT Press.

Słowiński, P., Alderisio, F., Zhai, C.m Shen, Y., Tino, P., Bortolon, C., Capdevielle, D., Cohen, L., Khoramshahi, M., Billard, A. Salesse, R., Gueugnon, M., Marin, A., Bardy, B.G., Bernardo, M., Raffard,S. and Tsaneva-Atanasova, K. (2017) 'Unravelling socio-motor biomarkers in schizophrenia', *NPJ Schizophrenia*, 3, 8. doi: 10.1038/s41537-016-0009-x

Tandon, R., Gaebel, W., Barch, D.M., Bustillo, J., Gur, R.E., Heckers, S., Malaspina, D., Owen, M.J., Schultz, S.K., Tsuang, M.T., Van Os, J., and Carpenter, W. (2013) 'Definition and description of schizophrenia in the DSM-5', *Schizophrenia Research*, 150 (1), pp. 3–10. doi: 10.1016/j.schres.2013.05.028

Targum, S.D., Schappi, J., Koutsouris, A. et al. (2022) 'A novel peripheral biomarker for depression and antidepressant response', *Molecular Psychiatry*, 27, pp. 1640–1646. doi: 10.1038/s41380-021-01399-1

Tellenbach, H. (1980) *Melancholy*. Pittsburgh: Duquesne University Press.

Thomson-Jones, M. (2012) 'Modelling without mathematics', *Philosophy of Science*, 79 (5), pp. 761–772. doi: 10.1086/667876

Toohey, M.J. and DiGiuseppe, R. (2017) 'Defining and measuring irritability: construct clarification and differentiation', *Clinical Psychology Review*, 53, pp. 93–108. doi: 10.1016/j.cpr.2017.01.009

Treadway, M.T., Buckholtz, J.W., Schwartzman, A.N., Lambert, W.E. and Zald, D.H. (2009) 'Worth the 'EEfRT'? The effort expenditure for rewards task as an objective measure of motivation and anhedonia', *PloS one*, 4 (8), e6598. doi: 10.1371/journal.pone.0006598

Trull, T.J. and Durrett, C.A. (2005) 'Categorical and dimensional models of personality disorder', *Annual Review of Clinical Psychology*, 1 (1), pp. 355–380. doi: 10.1146/annurev.clinpsy.1.102803.144009

Truran, P. (2013) 'Models: Useful but Not True', in Turan, P. *Practical Applications of the Philosophy of Science*. Berlin: Springer, pp. 1–14. doi: 10.1007/978-3-319-00452-5_10

Trzepacz, P.T. and Baker, R.W. (1993) *The Psychiatric Mental Status Examination*. Oxford: Oxford University Press.

Tversky, A. and Kahneman, D. (1974) 'Judgment under Uncertainty: Heuristics and Biases', *Science*, 185, pp. 1124–1131. doi: 10.1126/science.185.4157.1124

Tversky, A. and Kahneman, D. (1981) *Evidential impact of base rates*. Stanford: Stanford University, California Department of Psychology.

Ubl, B., Kuehner, C., Kirsch, P., Ruttorf, M., Diener, C. and Flor, H. (2015) 'Altered neural reward and loss processing and prediction error signalling in depression', *Social Cognitive and Affective Neuroscience*, 10 (8), pp. 1102–1112. doi: 10.1093/scan/nsu158

Upshur, R.E.G. and Colak, E. (2003) 'Argumentation and evidence', *Theoretical Medicine*, 24, pp. 283–299. doi: 10.1023/A:1026006801902

Varghese, F.T. and Mellsop, G.W. (1983) 'Formulating a psychiatric case', *Bulletin of the Royal College of Psychiatrists*, 7 (3), pp. 48–50. doi: 10.4103/0019-5545.58905

Verber, N. and Shaw, P.J. (2020) 'Biomarkers in amyotrophic lateral sclerosis: a review of new developments', *Current Opinion in Neurology*, 33 (5), pp. 662–668. doi: 10.1097/WCO.0000000000000854

Vertue, F.M. and Haig, B.D. (2008) 'An abductive perspective on clinical reasoning and case formulation', *Journal of Clinical Psychology*, 64 (9), pp. 1046–1068. doi:10.1002/jclp.20504

Volterra, V. (1926) Variazioni e fluttuazioni del numero d'individui in specie animali conviventi, *Accademia Nazionale dei Lincei*, 2, pp. 31–113.

Wagner, A.D., Pare-Blagoev, E.J., Clark, J. and Poldrack, R.A. (2001) 'Recovering meaning: left prefrontal cortex guides controlled semantic retrieval', *Neuron*, 31 (2), pp. 329–338. doi: 10.1016/s0896-6273(01)00359-2

Walter, H. (2013) 'The third wave of biological psychiatry', *Frontiers in Psychology*, 4, 582. doi: 10.3389/fpsyg.2013.00582

Walter, H., Daniels, A. and Wellan, S.A. (2021) 'Kognitive Neurowissenschaft des Positiven', *Nervenarzt*, 92, pp. 878–891. doi: 10.1007/s00115-021-01167-0

Warenski, L. (2021) 'Epistemic norms: truth conducive enough', *Synthese*, 198, pp. 2721–2741. doi: 10.1007/s11229-019-02242-5

Waters, C.K. (2007) 'Causes that make a difference', *Journal of Philosophy*, 104, pp. 551–579. doi: 10.5840/jphil2007104111

Waters, F., Blom, J.D., Dang-Vu, T.T., Cheyne, A.J., Alderson-Day, B., Woodruff, P. and Collerton, D. (2016) 'What Is the Link Between Hallucinations, Dreams, and Hypnagogic-Hypnopompic Experiences?' *Schizophrenia Bulletin*, 42 (5), pp. 1098–1109. doi: 10.1093/schbul/sbw076

Waters, F., Blom, J., Jardri, R., Hugdahl, K. and Sommer, I. (2018) 'Auditory hallucinations, not necessarily a hallmark of psychotic disorder', *Psychological Medicine*, 48 (4), pp. 529–536. doi: 10.1017/S0033291717002203

Waters, F., and Fernyhough, C. (2017) 'Hallucinations: A Systematic Review of Points of Similarity and Difference Across Diagnostic Classes', *Schizophrenia bulletin*, 43 (1), 32–43. doi: 10.1093/schbul/sbw132

Waters, F., Chiu, V., Atkinson, A. and Blom, J.D. (2018) 'Severe Sleep Deprivation Causes Hallucinations and a Gradual Progression Toward Psychosis with Increasing Time Awake', *Frontiers in Psychiatry*, 9, 303. doi: 10.3389/fpsyt.2018.00303

Watkins, E., Moulds, M., & Mackintosh, B. (2005) 'Comparisons between rumination and worry in a non-clinical population', *Behaviour research and therapy*, 43 (12), pp. 1577–1585. doi: 10.1016/j.brat.2004.11.008

Weerasekera, P. (1996) *Multiperspective case formulation: A step towards treatment integration*. Malabar: Krieger Publishing Company.

Weisberg, M. (2004) 'Qualitative theory and chemical explanation', *Philosophy of Science*, 71 (5), pp. 1071–1081. doi: 10.1086/428011

Weisberg, M. (2007). 'Who is a Modeler?', *The British Journal for the Philosophy of Science*, 58 (2), pp. 207–233. doi: 10.1093/bjps/axm011

Weld, D.S. and De Kleer, J. (ed.) (2013) *Readings in Qualitative Reasoning about Physical Systems*. Burlington: Morgan Kaufmann.

Wellan, W.H., Daniels, A. (2020) 'Glück im Unglück. (An)hedonie aus Sicht der kognitiven Neurowissenschaft und Psychiatrie', in Fink, H. and Rosenzweig, R. (ed.) *Hirn im Glück. Freude, Liebe, Hoffnung im Spiegel der Neurowissenschaft*. Nürnberg: Kortizes, pp. 67–90.

Wells, A. and Morrison, A.P. (1994) 'Qualitative dimensions of normal worry and normal obsessions: A comparative study', *Behaviour Research and Therapy*, 32, pp. 867–870. doi: 10.1016/0005-7967(94)90167-8

Wells, A. (1995) 'Meta-cognition and worry: a cognitive model of generalized anxiety disorder', *Behavioural and Cognitive Psychotherapy*, 23, pp. 301–320. doi: 10.1016/0005-7967(94)90167-8

Wells, A., and Carter, K. (1999) 'Preliminary tests of a cognitive model of generalized anxiety disorder' *Behaviour research and therapy*, 37 (6), pp. 585–594. doi: 10.1016/S0005-7894(01)80045-9

Westerkamp, D. (2006) *Via Negativa: Sprache und Methode der negativen Theologie*. Münster: Brill Fink.

Whooley, O. (2010) 'Diagnostic ambivalence: psychiatric workarounds and the Diagnostic and Statistical Manual of Mental Disorders', *Sociology of Health & Illness*, 32(3), pp. 452–469. doi: 10.1111/j.1467-9566.2010.01230.x

Whitaker, C.W. (1996) *Aristotle's De interpretatione: contradiction and dialectic*.Oxford: Oxford University Press.

Williams, J.B., Karg, R.S., Spitzer, R.L. and First, M.B. (2016) *User's Guide for the SCID-5-CV Structured Clinical Interview for DSM-5® Disorders: Clinical Version*. Washington: American Psychiatric Publishing, Inc.

Wimsatt, W.C. (1987) 'False models as means to truer theories' in Nitecki, M.H. and Hoffman, A. (ed.) *Neutral models in biology*. Oxford: Oxford University Press, pp. 23–55.

Woodward, J. (1984) 'A Theory of Singular Causal Explanation' *Erkenntnis*, 21, pp. 231–262. doi: 10.1007/BF00169275

Woodward, J. (2003). *Making Things Happen: A Theory of Causal Explanation*. Oxford: Oxford University Press.

World Health Organization (1993) *The ICD-10 classification of mental and behavioural disorders: diagnostic criteria for research*. Geneva: World Health Organization.

World Health Organization (2019) ICD-11: International classification of diseases. Accessible at: https://icd.who.int/ (Accessed: 01.09.2023).

Wright, B., Dave, S. and Dogra, N. (2017) *100 Cases in Psychiatry*. Boca Raton: CRC Press.

Xenakis, J. (1959) 'Plato's "Sophist": A Defense of Negative Expressions and a Doctrine of Sense and of Truth', *Phronesis*, 4 (1), pp. 29–43. doi: 10.1163/156852859x00056

Yager, J. (2020) 'Irritability Disorders in Adults: Diagnostic Categories Missing in Plain Sight?' *The Journal of Nervous and Mental Disease*, 208 (6), pp. 459–465. doi: 10.1097/NMD.0000000000001158

Ylikoski, P. (2007) 'The idea of contrastive Explanandum' in Persson, J. and Ylikoski, P. (ed.) *Rethinking explanation*. Berlin: Springer, pp. 27–42.

Ylikoski, P. and Kuorikoski, J. (2010) 'Dissecting explanatory power', *Philosophical Studies*, 148, pp. 201–219. doi: 10.1007/S11098-008-9324-Z

Ylikoski, P. (2013) 'Causal and Constitutive Explanation Compared', *Erkenntnis*, 78, pp. 277–297. doi: 10.1007/s10670-013-9513-9

Zahavi, D. (2019) *Phenomenology: The Basics*. Abingdon: Routledge.

Zetterqvist, M. (2015) 'The DSM-5 diagnosis of nonsuicidal self-injury disorder: a review of the empirical literature', *Child and Adolescent Psychiatry and Mental Health*, 9, 31. doi: 10.1186/s13034-015-0062-7

Zhang, W.N., Chang, S.H., Guo, L.Y., Zhang, K.L. and Wang, J. (2013) 'The neural correlates of reward-related processing in major depressive disorder: A meta-analysis of functional magnetic resonance imaging studies', *Journal of Affective Disorders*, 151, pp. 531–539. doi: 10.1016/j.jad.2013.06.039

Zimmerman, M., Ellison, W., Young, D., Chelminski, I. and Dalrymple, K. (2015) 'How many different ways do patients meet the diagnostic criteria for major depressive disorder?' *Comprehensive Psychiatry*, 56, pp. 29–34. doi: 10.1016/j.comppsych.2014.09.007